Offshore Semi-Submersible Platform Engineering

Offshore Semi-Submersible Platform Engineering

Srinivasan Chandrasekaran

CRC Press

Taylor & Francis Group

Boca Raton London New York

CRC Press is an imprint of the
Taylor & Francis Group, an **informa** business

First edition published 2021
by CRC Press
6000 Broken Sound Parkway NW, Suite 300, Boca Raton, FL 33487-2742

and by CRC Press
2 Park Square, Milton Park, Abingdon, Oxon, OX14 4RN

© 2021 Taylor & Francis Group, LLC

CRC Press is an imprint of Taylor & Francis Group, LLC

Library of Congress Cataloging-in-Publication Data

Names: Chandrasekaran, Srinivasan, author. Title: Offshore semi-submersible platform engineering / Srinivasan Chandrasekaran. Description: First edition. | Boca Raton, FL : CRC Press/Taylor & Francis Group, LLC, 2021. | Includes index. Identifiers: LCCN 2020032505 (print) | LCCN 2020032506 (ebook) | ISBN 9780367673307 (hardback) | ISBN 9781003130925 (ebook) Subjects: LCSH: Semi-submersible offshore structures–Design and construction. | Drilling platforms–Design and construction. | Structural analysis (Engineering) Classification: LCC TC1702 .C43 2021 (print) | LCC TC1702 (ebook) | DDC 627/.98–dc23

LC record available at https://lccn.loc.gov/2020032505
LC ebook record available at https://lccn.loc.gov/2020032506

ISBN: 978-0-367-67330-7 (hbk)
ISBN: 978-1-003-13092-5 (ebk)

Typeset in Times New Roman
by MPS Limited, Dehradun

Contents

Foreword by Anupam Gupta

In recent years, because of the increasing demand of stability and performance in the deep waters, semi-submersible platforms have gained significant popularity for various critical offshore projects such as drilling units, heavy lift cranes, and even hotel platforms (also known as flotels). The main reason behind the increased use of semi-submersibles lies in their advantages over other options, for example, higher payload capacity, better performance in unpredictable weather, and flexibility to relocate. Similarly, the demand of pipe-lay barges in challenging environments and deep waters has grown tremendously, leading to the requirement of optimized and efficient engineering methods.

As a practicing professional and researcher analyzing floating offshore structures, I have always felt the void of design handbooks explaining the latest analysis methods for special structures like semi-submersibles and pipe-lay barges. Considering the continuous developments in the field of offshore structures, it is crucial to have a reference always accessible to explain the basic concepts as well as the advanced research. This book fills the same gap providing a convenient tool for the design of semi-submersibles and pipe-lay vessels, with a novel perspective based on the study and research done by Prof. Chandrasekaran's research group. This book provides a detailed analysis of the specific offshore structures, semi-submersibles, and pipe-lay vessels, as well as an introduction to the offshore environment and the different types of floating platforms.

I am extremely delighted to write this foreword, not only because work by Dr. Chandrasekaran has inspired and guided me in my professional life, but also because this book provides a convenient guide for analysis of offshore structures. This book is written in the unique teaching style of Prof. Chandrasekaran, which has inspired me in my research and has benefited me personally in designing and analyzing offshore structures for critical oil and gas projects. Similar to the other books authored by Prof. Chandrasekaran, this book gives step-by-step explanations with examples describing concepts in an excellent style covering advanced topics of recent research, which makes it perfect for classroom materials, practicing engineers, and for budding researchers. For example, fatigue analysis, which is addressed in Chapter 3, provides a convenient solution for a complicated analysis in a simple and interesting fashion. In Chapter 3, Prof. Chandrasekaran has used an interesting example of a new generation compliant platform, Triceratops, to explain an advanced analysis procedure, providing MATLAB codes as an aid for readers with detailed flow charts. These examples not only help to clarify the concepts but also provide a template for handling design and analysis for critical projects.

I truly believe that this book will serve as an excellent reference for the researchers and engineers analyzing and studying complex behavior of the offshore semi-submersible platforms and pipe-lay barges.

Anupam Gupta, PE, B.Tech, MS, MBA, TechnipFMC

x

Foreword by Prof. Haluk Erol

To meet the need for offshore exploration and production of oil and gas, a new generation of semi-submersible platforms is being developed. Semi-submersible platforms are multi-legged floating structures with a large deck. Many of the semi-submersible platforms are unique in many respects, and their efficient and economic design and installation are a challenge to offshore engineering.

Offshore semi-submersible platforms should be designed to minimize both the excitation forces and the response of the structure to these excitation forces. A structure can be designed to optimize functional buoyancy forces, hydrostatic pressure, and natural frequencies to avoid resonances and fatigue and to minimize the excitation forces associated with wind, waves, currents, seismic events, and ice. The component member sizes of an offshore structure, including their configuration and arrangement directly influence the magnitude of these forces.

This book deals with the above-mentioned subjects: It discusses the excitation forces and the response of the offshore semi-submersible platform to these excitation forces in detail, presents both the theoretical and practical questions of the field, and gives a detailed description on the subject. It also shows the optimum design possibilities. Thus, it not only provides an analysis of the various offshore semi-submersible platforms but also presents guidelines for designers. The author has succeeded in summarizing the latest results from the field, and has published the results of his own observations regarding the techniques of optimum design of offshore semi-submersible platforms. The book will be a useful aid for engineers engaged in the design of offshore semi-submersible platforms as a handbook. It will expand their knowledge and assist them in choosing new solutions. The book can also be used as a textbook by universities and postgraduate students.

Prof. Haluk Erol

Foreword by Faisal Khan

Let me begin by introducing my background to make the context of my thoughts and assessment of the work presented in this book.

I am Professor and Canada Research Chair (Tier I) of Offshore Safety and Risk Engineering at Memorial University in Newfoundland, Canada. I established the Centre for Risk Integrity and Safety and Engineering (C-RISE), which has over 50 research members working on ocean and offshore safety, risk engineering, and risk-based integrity assessment and management. I work closely with marine and offshore energy industries. I have held many visiting positions, including but not limited to Risk Specialist at Lloyd's Register EMEA, UK; Professor of Ocean Engineering at Australia Maritime College, Australia; Professor of Offshore Safety at the China University of Petroleum, China; and Qatargas Chair Professor at Qatar University, Qatar. This provides me with the opportunity to teach, train, and facilitate the implementation of novel concepts of safety and asset integrity to the offshore system. I have authored over 500 research articles in peer-reviewed journals and conferences, and eight books on the subject area. This experience supports my expertise to share a broader and deeper perspective on the work presented in this book.

Professor Srinivasan Chandrasekaran needs no introduction. His research contributions and impact are well known; however, it is not the point of discussion here. I wish to highlight his exceptional contribution as an educator and trainer. Pof. Chandrasekaran has been active in the safety, risk, and reliability of the ocean and offshore system for quite some time. It is not an understatement to say he is the key expert on the subject area in the region and perhaps among a handful of experts globally. What makes Prof. Chandrasekaran's educational contribution unique is his ability to draw upon practical examples, simplify complex problems such that they are better understood before jumping to solutions, and straightforwardly presenting novel ideas. While safety, risk, and reliability topics are often studied using mixed qualitative and quantitative approaches, Prof. Chandrasekaran's focus has remained on quantitative methods, which are repeatable, auditable, and easily implementable. Prof Chandrasekaran communicates complex quantitative approaches effectively. One such example is the present book. Before continuing, I wish to reiterate that Prof. Chandrasekaran has discussed concepts with numerical examples, enabling this book to serve as a primary learning source.

The topic of Offshore Semi-Submersible Platforms is broad, while the content presented here is quite deep. The work draws upon practical examples that Prof. Chandrasekaran and his team have been working on for quite some time. Semi-submersible platforms are explored as a venue to support energy production in deeper and remote waters. While this provides economic benefits, the most pertinent questions that remain unanswered are "What are the safety and asset integrity of this development?" "Are these developments more vulnerable or more resilient?" and the more intriguing question is, "What are the impacts of climate change (harsh environment) on the safety and integrity of operations?" The ocean engineering community must guide the current development to make it safer and sustainable.

This book touches upon this theme. It provides foundational knowledge on safety, reliability, resilience, and integrity management of the offshore system. Integrating the knowledge presented here with the experience brings us a step closer to developing innovative solutions to prevent failures, incidents, and accidents. This helps to improve safety, integrity, and resilience of the offshore system. This book serves a vital role in this regard.

The book comprises five chapters, each dedicated to the core issue of offshore systems. The climate change is redefining the loads; a rare event becoming not so rare and thus demanding reconsideration of load estimation. Chapters 3 and 4 covered this topic to a great extent. These chapters provide readers with a good pointer on how to begin such an analysis. Chapter 5 (offshore pipeline) is a topic of greater importance to the offshore energy industry, given its practical relevance and growing use. Prof Chandrasekaran has presented some foundation knowledge on this topic, which will help academics and practitioners.

In summary, I conclude this book serves a knowledge bank on the offshore system's focused topic. It does not address all issues we are facing; however, it does identify critical challenges and provide foundational knowledge to understand better and analyze these challenges. I am confident readers will enjoy and greatly benefit from using this book.

Faisal Khan, PhD, P.Eng
Canada Research Chair (Tier I) in Offshore Safety & Risk Engineering
Associate Dean (Graduate Studies)
Director, Centre for Risk Integrity and Safety Engineering (C-RISE)
Faculty of Engineering & Applied Science
Memorial University, St John's, Canada

Preface

The enormous petroleum and natural gas reserves in the ocean have not been thoroughly explored due to the hostility and harsh weather conditions prevailing in these places. The ancient structures constructed for these purposes were similar to land-based structures and were rigidly fixed to the seabed. These offshore constructions attracted large environmental loads, which led to the failure of the entire system itself. To overcome this issue, engineers prefer truss-like structures, which allowed passage of waves encountering them. But when the explorations begin to expand toward the deep sea, these structures also became old due to economic and technical limitations. Easy construction, handling, and installation of floating structures gained popularity among offshore technologists. The challenge in simulating the real sea state of a particular site fueled scientists to introduce radical analytical and numerical solutions. Floating structures need to remain stable while in operation. Offshore platforms like semi-submersibles should be position-restrained for purposeful functioning. The necessary restoration is provided by buoyancy or mooring lines. Operations of semi-submersibles impose many challenges as they bring a high degree of nonlinearity in their behavior. These challenges demand a detailed analysis of the platform vessel motion and the proper choice of the mooring layout. Environmental loads acting on the platform induce cyclic loads, which have the potential to cause fatigue stresses over a long period. It is essential to understand that the re-centering of the floating platforms is fundamentally due to the mooring lines and their failure may affect the platform stability.

A postulated failure analysis of mooring lines under the worst combination of environmental loads is necessary to examine the reliability of the platform. The author attempts to present a detailed analysis and design procedure to this objective through studies on the semi-submersible platform. Mooring line layouts of the semi-submersible and pipe-laying barges based on detailed analyses will serve as a reference for designers, graduate students, and researchers. The author acknowledges the contributions of his research team members: Syed Azeem Uddin, Shihas Khadhir, and Nagavinothini, whose detailed studies have helped us to arrive at useful discussions presented in the chapters of this book. The author also thanks the Center of Continuing Education, Indian Institute of Technology Madras, for its administrative support extended during the book writing. Whole hearted thanks are due to all the researchers who have contributed to the subject content of this domain in the past. Numerical analyses and inferences drawn from the example studies are only indicative and need to be verified by the user when adopting them for practical design.

Srinivasan Chandrasekaran

MATLAB® is a registered trademark of The MathWorks, Inc. For product information, please contact:

The MathWorks, Inc.
3 Apple Hill Drive
Natick, MA 01760-2098 USA
Tel: 508-647-7000
Fax: 508-647-7001
E-mail: info@mathworks.com
Web: www.mathworks.com

Author

Srinivasan Chandrasekaran is a full professor of structural engineering in the Department of Ocean Engineering, Indian Institute of Technology Madras. He commands a teaching experience of 28 years, during which he has guided 30 research theses. He has authored 15 textbooks, published by publishers of international repute. He has also published 160 research papers in peer-reviewed international journals and refereed conferences. He holds domain expertise in a few areas, namely: structural dynamics, reliability and risk assessment, ocean structures and materials, advanced structural analysis, and HSE in offshore and petroleum engineering. More details can be found at: http://www.doe.iitm.ac.in/drsekaran.

1 Introduction

1.1 OCEAN ENVIRONMENT

The ocean environment is highly complex as it generates various types of loads upon the structures constructed in the sea. However, wave-structure interaction, which dominates the geometric design of offshore structures, is a well-known and clearly understood phenomenon. Wave-structure interaction has been successfully taken into consideration in the design principles of ships, and then later extended to the design of offshore structures. While conventional design of structures deals with gravity loads in static design and lateral loads in the dynamic model, load combinations that act on offshore structures are complex and novel. A large variety of environmental loads that act on offshore structures include wave loads, wind loads, current, ice loads, and impact loads. Several complexities that are present in the ocean environment make these load computations highly mathematical and uncertain. As geographical phenomena control the amplitude and period of these loads, say, for example, wave loads, they are estimated using different idealized theories and empirical relationships.

1.2 ENVIRONMENTAL LOADS

Environmental loads continuously interact with offshore structures in the form of waves, wind, current, or earthquakes. Ocean structures are repeatedly loaded, and these cyclic loads are capable of inducing fatigue damage to the structure over time. Among all these loads, two that are of primary concern are wave and wind loads; the former is a high-frequency phenomenon in comparison with the latter. Offshore structures are also prone to seismic loads when located near tectonic zones (Sigurdsson, 1988). Unlike the conventional design procedures, loads that act on offshore structures should be assessed in challenging the strength of the structural members (Chandrasekaran and Ajesh, 2019); many of the loads pose a challenge to the structural stability. This is because they act while the structure is in a floating condition. Most of the compliant structures have inherent stability only through their geometry and not by the material and member strength.

1.2.1 WAVE LOADS

The most important of all environmental loads is the wave load. Waves play a critical role in the design of offshore compliant structures due to the complications involved in the hydrodynamic behavior of the platform in open sea conditions. Wave analysis can be performed either by the design wave concept or the statistical approach. In the design wave concept, a regular wave is defined using the wave height (H) and the wave period (T). The wave forces are then calculated using the appropriate wave theories. Water particle kinematics are calculated as a function of

1

sea-surface elevation using the potential theory; waves are assumed to be long-crested. Various wave theories are developed to include a wide range of wave parameters; the most common theory, which is widely used, is Airy's linear wave theory. In the statistical approach, random waves are generated using the appropriate wave spectra, which is site-specific. The most probable maximum wave force is then computed using the linear wave theory. A statistical approach is highly essential to accurately assess the dynamic behavior and the fatigue strength of the structure. It helps arrive at a response spectrum that defines the maximum expected response within a particular interval of time.

The motion of the water particles induces wave loads on the offshore structures, which, in turn, results in their dynamic behavior. Wave loads on offshore structural members are computed based on the size of the structural members encountering the wave loads. Slender members with a diameter (D) to wavelength (L) ratio of less than 0.2 will not influence the wavefield; therefore, wave loads can be calculated using Morison's equation. But, large-diameter members interfere with the wavefield; in such cases, wave loads are calculated using diffraction theory. Generally, ocean waves are random but can be represented as a regular wave described by a deterministic approach.

As the waveform in each cycle in a regular wave is the same, wave theories describe the characteristics of a typical cycle; it remains invariant for other cycles. Two significant parameters are the period (T) and height (H) of the waves and water depth (d). A sample time history of a regular wave is shown in Figure 1.1.

Based on the structural waveform, waves can be classified as regular waves and irregular or random waves. Initially, the transfer of energy causes capillary waves, which grow to form irregular waves with different amplitudes and periods depending upon the wind speed and direction. If the wind blowing over the ocean surface has a constant wind velocity, then the generated irregular wavelets will grow to a fully developed sea with constant wave amplitude and period. Such waves are referred to as regular waves, and these waves exhibit a sinusoidal motion.

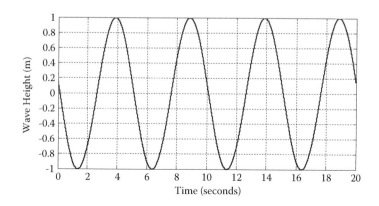

FIGURE 1.1   Typical regular wave (2 m, 5 s).

Because the actual wave formation is complex, the following mathematical formulations are always used to model the ocean surface with regular waves.

1. Airy's two-dimensional small-amplitude linear wave theory
2. Stokes theory
3. Solitary wave theory
4. Cnoidal theory
5. Stream function theory

Airy's theory, which assumes linearity between the wave height and kinematic quantities, is commonly used in the theories listed above. The regular waves are usually defined by their wave height (H) and wave period (T), as shown in Figure 1.2. For determining the wave forces on offshore structures, the wave surface profile should be idealized. Any one of the above mentioned appropriate wave theories should be used to compute the water particle kinematics. The applicability of wave theory is based on a wide range of parameters, including the wave height, water depth, and wave period. It is not always possible to select the wave theory precisely suitable for the selected condition. Airy's small-amplitude linear wave theory is valid for deep water conditions where $(d/gT^2) > 0.8$, and Stokes theory should be used when $(H/gT^2) > 0.04$.

The most straightforward wave theory is Airy's linear wave theory, or small-amplitude wave theory. According to this theory, the waveform has a sinusoidal profile. This theory also provides the kinematic and dynamic amplitudes as a linear function of wave amplitude or wave height. Thus, the normalized amplitude value is unique and invariant to the wave amplitude. It helps to represent the response of the offshore structures as a normalized value. The normalized responses, as a function of wave height, are called the transfer function or the Response Amplitude Operator (RAO). This method is simple and predicts the extreme response of the structures.

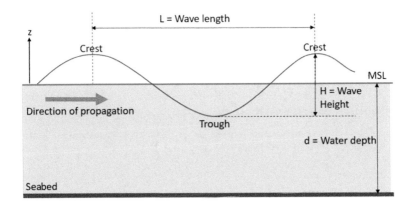

FIGURE 1.2 Wave parameters.

However, in reality, ocean waves are a combination of a set of waves with different frequencies and directions; they appear as irregular or random waves (Chandrasekaran and Anubhab, 2004; 2005). Random waves are represented by wave energy density spectra, which describe the ocean wave's energy content. It is spread over a wide frequency ranging from zero to infinite value, but waves are found to be concentrated on a narrow band. Their statistical parameters characterize random waves. Sea states are represented by the significant wave height (H_s) and zero-crossing periods (T_z). Different sea states used to describe random waves are summarized in Table 1.1.

Several spectral models are available for use in the design of offshore structures. These models are derived based on the observed properties of the ocean and are empirical. The frequency characteristics of the real sea conditions influence the spectral formulation. The most commonly used spectral formulas include the P-M spectrum, JONSWAP spectrum, ISSC spectrum, Bretschneider spectrum, and Ochi-Hubble spectrum (Chandrasekaran, 2015a,b). Each spectrum model distributes the wave energy differently across the frequency band, so the structure's response will vary for the same wave height if different spectra are used. The commonly used spectral models in offshore structural design are the Pierson Moscowitz (P-M) spectrum and the JONSWAP spectrum. The P-M spectrum applies to various regions such as the Gulf of Mexico, Offshore Brazil, Western Australia, Offshore Newfoundland, and Western Africa, both in operational and survival conditions. The JONSWAP spectrum applies only to the North Sea for operational and survival conditions. The P-M spectrum, which represents the wave energy distribution under different frequencies, is suitable for open sea conditions. This spectrum is neither fetch limited nor duration limited. It is a two parameter spectrum, developed under moderate wind conditions existing over large fetches and is given by the following relationship:

$$S^+(\omega) = \frac{1}{2\pi} \frac{H_s^2}{4\pi T_z^2} \left(\frac{2\pi}{\omega}\right)^2 \exp\left(-\frac{1}{\pi T_z^4}\left(\frac{2\pi}{\omega}\right)^4\right) \quad (1.1)$$

where H_s is the significant wave height, T_z is the zero-crossing period, and ω is the frequency. The spectral plot shows that the wave energy is concentrated on a

---

TABLE 1.1

Characteristics of random sea states

Sea state description	Significant wave height H_s (m)	Zero-crossing period T_z (s)	Wind velocity (m/s)
Moderate	6.5	8.15	15
High	10	10	35
Very high	15	15	45

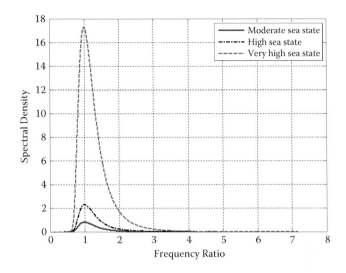

FIGURE 1.3 P-M spectrum for different sea states.

narrow band. The typical wave energy P-M spectra under different sea states are shown in Figure 1.3.

The sea state can be described using a suitable scale based on sea conditions, wave height, and wind speed. One such empirical measure used for expressing the sea state is the Beaufort scale, developed by Irish Royal Navy Officer Francis Beaufort in the 19th century. As per the World Meteorological Organization, the scale is standardized into thirteen classes, from zero to twelve. Because the description of the sea surface is more useful than the wind speed and wave heights alone, the World Meteorological Organization and MetService started defining the sea state using the Douglas sea scale, developed in 1920 by Captain H.P. Douglas. It describes the roughness of the sea from smooth to very high. However, offshore structures are also rehabilitated using a perforated cover around the main members to reduce the response from the wave impact (Chandrasekaran and Merin, 2016; Chandrasekaran and Madhavi, 2014a–e; 2015a–d; 2016; Chandrasekaran et al., 2013b; 2014e; Chandrasekaran and Abhishek, 2010).

1.2.2 WIND AND CURRENT

In addition to wave loads, offshore platforms are also subjected to wind and current loads. Offshore complaint platforms are designed to remain flexible in the horizontal plane. As a result, they have significant surge, sway, and yaw periods (60 to 150 s), granting them freedom for large displacements under lateral loads. These motions are of low frequency and get more excited in the presence of wind. The fluctuating wind component, called the gust component, induces low-frequency oscillations. The magnitude of oscillation increases with the increase in the platform's exposed area and the wind velocity. Another common occurrence introduced by wind effect in the open sea is current. Current adds varying pressure distributions around the structural member, resulting in a steady drag force. The combined action of wind, waves, and current introduces a significant response in the compliant offshore platforms.

A significant source of wave generation is the wind; hence, the analysis of the structure in the wave-only environment is not realistic. In addition to wind-induced waves, wind also generates load on the superstructure. The dynamic wind effect will be significant on the compliant offshore platforms. Therefore, the impact of both mean wind and the gust component should be considered in the analysis. In the design of offshore structures, an average wind speed occurring over a one-hour duration is taken as a steady wind speed. It is typically measured at 10 m above the mean sea level (MSL). Wind loads acting on the deck causes offset, resulting in the set-down of the deck as well. It induces additional moment in pitch degree-of-freedom, which influences tension variation in the tethers or the mooring lines. Thus, a strong coupling is developed between the surge, heave, and pitch, which is inherent in compliant structures.

A wind spectrum can describe random wind blowing over a structure. Various spectral formulations such as Davenport (1961), Harris (1971), Kaimal et al. (1972), Simiu and Scanlan (1978), and the American Petroleum Institute (1989; 2007) spectra show a significant difference at lower frequencies in the spectral density plots. The Davenport spectrum has more moderate energy content at lower frequencies than other wind spectra (Zaheer and Islam, 2008; 2012; 2017). Besides, the Davenport spectrum, developed for land-based conditions, may not represent the wind velocity fluctuations at low frequencies in the offshore environment. Therefore, the most preferred wind spectrum for the analysis of offshore structures is the API spectrum. It shows a higher energy content in the lower frequencies compared to other spectral formulations. The equation for the API spectrum is given by:

$$\frac{\omega S_u^+(\omega)}{\sigma_u(z)^2} = \frac{\theta}{(1 + 1.5\theta)^{5/3}} \qquad (1.2)$$

where θ is the frequency ratio or derivable variable $\left[\theta = \frac{\omega}{\omega_p}\right]$, ω_p is the peak frequency, Z_s is the surface height (20 m), $\sigma_u(z)^2$ is the variance of U(t) at a reference height, z is the reference height (= 10 m), and $S_u^+(\omega)$ is the spectral density (Chandrasekaran, 2014; 2016; 2017; 2019a,b). Variance $\sigma_u(z)^2$ at the reference height is given by:

$$0.01 \leq \frac{\omega_p^2}{\bar{U}_z} \leq 0.1 \qquad (1.3)$$

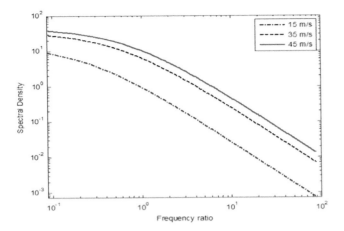

FIGURE 1.4   Spectra plot for different wind velocities.

$$\sigma_u(z) = \begin{cases} 0.15\bar{U}_z \left(\frac{z_s}{z}\right)^{0.125} & (if \to 2 \le z_s) \\ 0.15\bar{U}_z \left(\frac{z_s}{z}\right)^{0.275} & (if \to 2 > z_s) \end{cases} \qquad (1.4)$$

The API wind spectral density plot for different wind velocities is shown in Figure 1.4, which shows a significant energy concentration at lower frequencies. Unlike the wave spectra, the wind spectra are found to be broad-banded. Current occurs commonly in the open sea, due to the wind effect on water, tidal motion, temperature differences, density gradients, and salinity variations. The presence of current modifies the water particle kinematics and the apparent wave period. It also imposes drag forces on structures. The current velocity varies with depth; the highest value is observed near MSL. The wind-generated current is applied in the same direction as that of the wave and wind loads.

Wind is the air circulation in the atmosphere, occurring due to unequal heating of the earth's surface between the tropics and poles. Wind movement occurs due to the transfer of heat energy, which is profoundly affected by the rotation of the earth. The deflection of wind due to the rotation of the earth is called the Coriolis effect. The wind is responsible for the generation of ocean waves and currents. The wind loads on offshore structures are quite complex due to the mean component and time-varying component of wind velocity, as shown in Figure 1.5. The time-varying component contributes to the dynamic effect of wind load on the structures. However, structures are designed by considering the wind loads as static by incorporating the dynamic component indirectly using the gust factor, which is the currently accepted method for the computation of wind effects on the structure. The time-varying component of wind velocity is termed wind gust, and it depends upon the mean wind velocity, the height from the ground level, and the surface roughness of the exposed area. The wind direction depends upon the pressure gradient and the

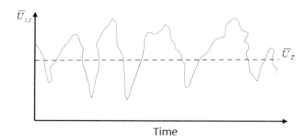

FIGURE 1.5 Dynamic variation of wind velocity.

Coriolis force. The wind velocity is constant at greater heights and is referred to as the gradient wind speed. At lower altitudes, eddy formation due to the surface roughness of earth causes the wind gusts to vary continuously with space and time. Wind speeds of above 20.0 m/s and a fetch length greater than 30.0 km at 10.0 m above the sea surface is assumed to be a stable wind condition, where the wind time history is stationary.

The wind is usually separated into mean and dynamic components for analytical considerations. The mean wind velocity and associated dynamic components are always calculated over a particular period due to the variation of the wind velocity concerning space and time. From the spectral analysis of wind velocity time history over such an extended period, the energy spectrum of wind can be developed. The wind spectrum consists of short period components due to gust, while the long-period component occurs due to the change in pressure. There is a spectral gap of about 360 s to 1.60 hours. It results from the lack of contribution from both wind gusts and pressure change. The pressure effect and the wind gusts change the wind velocity slowly and rapidly about the mean, respectively. An hour-average wind period is used for analysis, and the design wind speeds are usually specified by considering its return period. The commonly used wind spectra for offshore locations are the Kaimal spectrum and American Petroleum Institute (API) spectrum. The spectral density plot for a mean wind speed of 15 m/s and a period of 10 seconds is shown in Figure 1.6.

The more straightforward equation used to represent the distribution of wind speed to height is the power law distribution, which is given by

$$\bar{U}_Z = \bar{U}_{10}\left(\frac{z}{10}\right)^{1/7} \tag{1.5}$$

$\bar{U}_{10}$ is the basic design wind speed at 10.0 m above the Mean Sea Level (MSL), $\bar{U}_Z$ is the wind speed at height z from the MSL (API, 2000). The spatial variation of mean wind velocity in an offshore location is shown in Figure 1.7. It can be seen from the figure that the wind load acts on the exterior and transparent portions of the deck of the platform.

From the design wind velocity, the wind force acting horizontally on a surface can be calculated using the following equation:

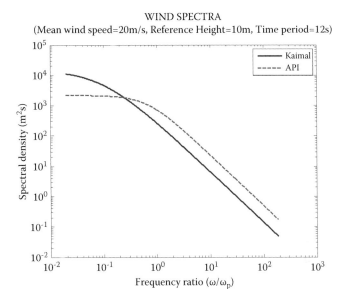

FIGURE 1.6   Wind spectra applicable to offshore locations.

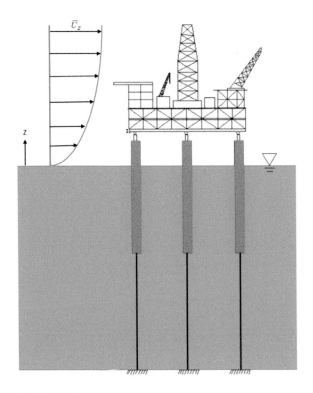

FIGURE 1.7   Spatial variation of mean wind velocity.

$$F_H = \frac{1}{2}C_D\rho_a U_z^2 A_p \qquad (1.6)$$

where F_H is the horizontal wind force, C_D is the drag coefficient, ρ_a is the density of wind, and A_p is the projected area. The value of the drag coefficient depends upon the shape of the sections under consideration and the wind velocity. In the case of an offshore deck, the wind force should be calculated for all the members separately and then added together.

Current motion occurs due to the transfer of water mass from one place to another, which generally occurs due to the wind forces, temperature difference, tides, change in density, and salinity and river discharge. The landmasses, continental shelves, and the rotation of the earth also modify the ocean currents. The action of the current results in drag forces on offshore structures and thus modifies its response. Tidal current velocity is similar to the sinusoidal motion of ocean waves with long periods, and the maximum tidal current occurs near mid-tide. Extreme wind conditions lead to the formation of surface wind drag currents along with the waves. However, the direction of current does not always acts in the same direction of the wave and current. But, for engineering conservative design purposes, the wind-generated currents are assumed to move in the same direction as the waves. The wind-generated current velocity variation concerning depth is considered to be linear for design purposes, varying from 1.0% to 3.0% of sustained wind speed at MSL to zero at the sea bottom. Typical tidal current and wind drag current profiles are shown in Figure 1.8. It should be noted that the wave-current interactions also govern the design of offshore platforms in ultra-deep-waters because the current acting along the direction of the wave increases the wavelength. The current velocity should be added vectorially to the water particle velocity in the drag force component while computing the wave force using Morison's equation.

1.2.3 ICE LOADS

In the recent past, there has been a significant increase in oil drilling in the Arctic region and other ice-infested waters in ultra-deep-water conditions. In addition to

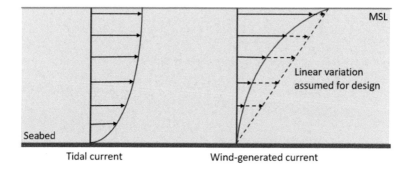

Tidal current   Wind-generated current

FIGURE 1.8   Current velocity variation.

the hindrance caused to normal operations, ice sheets also induce dynamic effects on offshore structures. Oil and gas exploration in Arctic regions also pose design challenges due to their cold environmental conditions. Offshore platforms, commissioned in ice-infested water, need to be assessed for the hindrance caused by ice sheets to their operability and the nature of random and cyclic characteristics of ice loads that impose dynamic effects (Shih, 1991).

Drifting of ice occurs due to several environmental factors such as wave, wind, or current, which highly depend upon the location and sea state. In offshore engineering, it is commonly assumed that the ice loads are influenced by the limit-stress due to the availability of sufficient energy to cause ice failure. Thus, the ice force is that force required causing failure of ice within the vicinity of the structure. One of the most critical factors that limit the maximum ice load acting on the structure is the ice-failure mechanism. It, in turn, depends on several factors such as ice velocity, width, and shape of the structure. Failure may occur by crushing, splitting, bending, buckling, and a combination of modes (Chandrasekaran and Nagavinothini, 2020a–d; 2019a–c). Several researchers reported that the ice sheet's crushing failure tends to cause maximum ice force on the offshore structures (Yue et al., 2001; 2002; 2009). Crushing ice failure occurs when a thick sheet of ice with a moderate velocity impinges the vertical side of any member. Ice-structure interaction causes a horizontal crack on the ice sheets, which results in the pulverization of ice. The crushed and pulverized ice sheet piles up over the intact ice sheet, which is supported by the buoyancy force. It causes an upward and downward motion of the ice edge adjacent to the offshore structure, inducing vibration to the structure. The total ice force applied to the structure under crushing failure is a function of ice strength in crushing.

When ice interacts with a compliant structure, the failure modes are ductile and brittle at low and high indentation rates, respectively. Continuous, brittle crushing at the top indentation rate results in partial contact and non-uniform pressure. Ice-force time histories have either a constant amplitude periodic waveform or randomly distributed periods and amplitudes. Under very high ice velocity, the ice force can be designated as a stochastic process. In such cases, ice forces can be described using a frequency spectrum. As the transition between the different modes of failure under crushing is not entirely known, the uncoupled, time-dependent ice load can be used in the dynamic analysis of offshore and polar structures.

Mechanical properties of sea ice are highly influenced by temperature, as sea ice exists very close to its melting temperature and also consists of brine. Several expressions have been developed for determining the ice strength, from the strain rate and porosity for crushing failure of ice. The ice crushing force is the product of the crushing pressure and the contact area. Effective ice pressure depends on several factors, namely, aspect ratio, confinement within the ice sheet, scale, and degree of contact between the ice and structure (Sodhi and Haehnel, 2003). The ice load limit can be computed using Korzhavin equation (Reddy and Swamidas, 2016) for ice crushing against the vertical structure and is given by the following expression:

$$F = a_1 a_2 a_3 h w \sigma_c \qquad (1.7)$$

where a_1 is the shape factor (0.9 for circular members), a_2 is the contact factor (0.5 for moving ice), a_3 is the aspect ratio factor, σ_c is the crushing strength of ice in MPa, h is the thickness of ice in meters, and w is the projected width of the structure in meters.

The crushing strength of ice depends on the temperature. The maximum ice-crushing strength in the coldest time of the year, as recorded in the Beaufort Sea, is 3 MPa, which can be considered an extreme value. Under spring conditions, with the temperature closer to the melting point, ice-crushing strength reduces to half of its value, which is accorded as the normal condition. Though Korzhavin's equation is one of the fundamental approaches, it is still in use with modification factors. Local thickening of ice is also taken into consideration by contact factor a_2. However, it should be noted that the local thickening occurs only under a stationary ice load with prolonged contact. Thus, the contact factor is chosen as 0.5 for moving ice. The previous studies also reported that the limit force obtained from Korzhavin's equation is comparable with the empirical relations derived for the prediction of limit ice force at lower ice thickness. Further, the questions also exist with the applicability of other empirical formulations for the prediction of ice force in compliant multi-legged offshore structures (McCoy et al., 2014).

Though there are several ice failure modes, crushing failure causes the maximum force in level-ice action. Continuous ice-crushing occurs under the high speed of ice movement. When the drifting ice interacts with the structure, a horizontal crack occurs at the edge of the ice sheet. Further, flaking divides the ice sheet into several layers. A portion of the ice sheet will pile up and slide around the ice surface due to crushing and pulverization. Under brittle crushing, the ice-crushing strength will not influence the response as the crushing strength becomes constant. Under such conditions, the feedback mechanism induced by the compliant structure will become negligible. A spectral model for ice speed ranging from 0.04 m/s to 0.35 m/s, suggested by Karna et al. (2007) has a non-dimensional spectral density function as given below:

$$\bar{G}_n(f) = \frac{af}{1 + k_s a^{1.5}f^2} \tag{1.8}$$

where $a = bv^{-0.6}$, v is the ice velocity in m/s, b, and k_s are the experimental parameters, and f is the frequency in Hertz. The following expression gives the spectral density function:

$$G_n(f) = \frac{\sigma_n^2 \bar{G}_n(f)}{f} \tag{1.9}$$

where σ_n^2 is the variance of the local force. The mean ice force and standard deviation are calculated from the following set of equations:

$$\sigma_n = \frac{I_n}{1 + kI_n}F_n^{max} \tag{1.10}$$

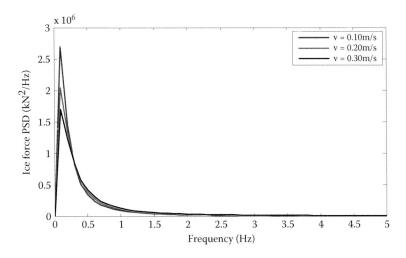

FIGURE 1.9 Spectral density for different ice velocities.

$$F_n^{mean} = \frac{F_n^{max}}{1 + kI_n} \tag{1.11}$$

where I_n is an intensity parameter, which varies from 0.2 to 0.5; k is the selected probability of exceedance; F^{mean} is the mean ice force; and F^{max} is the maximum ice force. The spectral density plot developed from the above set of equations for an ice thickness of 1.5 m and 1.5 MPa ice crushing strength is shown in Figure 1.9. It is also seen that the peak of the spectrum varies according to the ice velocity; maximum energy is found at the lower frequencies. The variation in the spectrum due to the ice velocity is seen at frequencies lesser than 0.3 Hz. Thus, it can be predicted that the effect of the ice velocity on the time-varying component on the ice force is less, and the factor that dominates the structure's response is the mean force. Several factors also influence ice loads on multi-legged offshore structures. These include interference and shielding effects between the legs and jamming of ice between legs. It depends upon the arrangement of legs and the clear spacing between the legs. For example, in the case of a three-legged structure, the maximum load occurs when the ice approaches the two legs, directly. In such cases, the maximum total load is twice that of a single leg.

As the ice spectrum model is developed based on the real-time data collected from offshore structures, the model automatically accounts for the formation of ice prows and ice rubble mounds during ice structure interaction. A significant variation in the spectral energy is observed with the change in the maximum ice force acting on the structure, as shown in Figure 1.10, for 0.2 m/s ice velocity. The plot shows that the response of the structure will be profoundly affected by the mean ice force. A time series of the ice force is then developed from the spectrum for different ice load cases using the inverse fast Fourier transform; sampling frequency is taken as 20 Hz, and the coherence function is evaluated. Total ice force is given by the sum

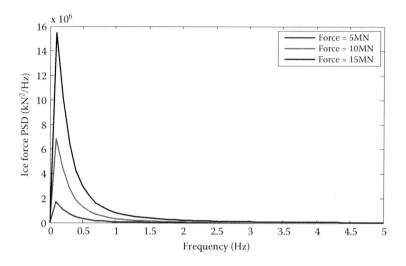

FIGURE 1.10   Spectral density under different ice forces.

of the mean ice force and the time varying ice force. The following expression gives the spectral density function of the total force:

$$G_{FT}(f) = (C + \mu_K S)^T G_{FF}(f)(C + \mu_K S) \tag{1.12}$$

where C is used for the consideration of the angle of incidence of the local forces and S is used to consider the local shear forces. The external global pressure acting on the structure can be found out by obtaining the standard deviation and the maximum peak value from the above equation. The ice load time history obtained for 0.5 m ice thickness, 1.5 MPa ice crushing strength, and 0.2 m/s ice velocity are shown in Figure 1.11.

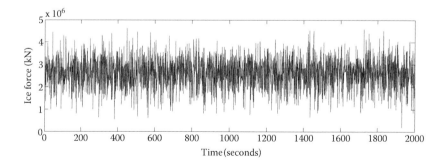

FIGURE 1.11   Ice force-time history.

FIGURE 1.12   Fire triangle.

1.2.4 FIRE LOADS

Fire is a chemical reaction involving rapid oxidation or combustion of a fuel. The swift reaction of vapor and oxygen initiated with the formation of the flammable mixture by heating is called the combustion process. The complete combustion of hydrocarbons forms water and carbon dioxide. The incomplete combustion will result in the formation of toxic byproducts and dense black smoke. Ignition is the self-sustaining combustion that may occur due to the rise in bulk temperature of the flammable mixture. The temperature at which this phenomenon occurs is called the autoignition temperature, which varies determined by each type of gas and liquid. Thus, fire is an exothermic oxidation-reduction reaction that usually involves oxygen. Therefore, three necessary conditions, fuel, oxygen, and heat, are required for combustion to take place. It is represented by a fire triangle shown in Figure 1.12. It also indicates that all three components should be present for the fire to occur (Chandrasekaran, 2016b). Thus, to avoid a fire in petrochemical facilities, the escape of fuel should be stopped, or heat should be removed, or the supply of oxygen should be stopped.

The heat generated through fire is transferred by conduction in solids, convection in fluid, and radiation, which does not require a medium. The transfer of heat to offshore structural components and equipment can lead to failure of the structure. Conduction is the primary mode of heat transfer and occurs through the motion of the molecules and migration of free electrons, mainly associated with the heat transfer in a material. Conduction (Co) is described using Fourier's law as follows:

$$Co = -kAT \qquad (1.13)$$

where T is the temperature gradient of the solid; A is the cross-sectional area of the solid; and k is the thermal conductivity of the material, which is the measure of the material's ability to transfer heat. Thus, materials with low heat conduction should be used for insulation purposes. The motion of the fluid under a change in temperature is described using convection. The convection process, which transfers the energy from the liquid to the adjacent solid, also increases the rate of heat transfer by conduction. The following relationship gives the rate of convection heat transferred (C_r), as per Newton's law of cooling:

$$C_r = -hA\left(T_{surf} - T_{amb}\right) \qquad (1.14)$$

where h is the convective heat transfer coefficient, A is the area of the contact surface, T_{surf} is the hot surface temperature, and T_{amb} is the ambient fluid temperature. Radiation is the transmission of electromagnetic energy through space, which transfers the heat between two substances that are not in contact with each other. The radiation heat depends upon the temperature of the structure and the surrounding environment. The Stefan-Boltzmann law is used for describing the rate of transfer of radiant thermal energy:

$$RF = esT_e^4 \qquad (1.15)$$

where RF is the emissive power flux of the surface, s is the Stefan-Boltzmann constant, e is the emissivity of the surface, and T_e is the emitter's temperature. The following expression gives the rate of heat transfer between two surfaces (a, b):

$$R = es(T_e a^4 - T_e b^4) \qquad (1.16)$$

The radiation heat transfer is dominant in the case of many hydrocarbon fires, where the heat radiates from the flame with soot. It may also result in severe burns to human beings and even heat the unprotected structural components and equipment considerably.

Offshore oil drilling and production platforms handle a large volume of flammable and combustible materials, and the probability of ignition is very high during the processing and storage of petrochemicals. Based on the type of flammable material, mechanism of release, temperature, and ignition point, different types of fire scenarios may happen. It includes Jet fire, Pool fire, Flashfire, Running liquid fire, Fireball or Boiling Liquid Expanding Vapor Explosion (BLEVE), and Vapor cloud explosions (VCEs).

The delayed ignition of a suitable quantity of flammable material with adequate mixing of gas or vapor and air in a confined area is called a Vapor cloud explosion. When the space is unconfined, it will result in a flash fire due to the burning of a vapor cloud of flammable materials. The duration of the flash fire is concise. Increased and consistent release of combustible material may convert the flash fire into the jet fire. Jet fire usually occurs due to the accidental release and burning of gas at flares. Small openings and high-pressure release characterize these. The sudden release of pressurized liquid or gas will result in a fireball, typically of 5 to 20 seconds duration. Fireball occurs due to Boiling Liquid Expanding Vapor Explosion (BLEVE). Though the heat loads from fireball do not affect the structure, the impact of BLEVE can be significant. The release of flammable material from the process equipment causes pool fires. The heat transferred from a pool fire depends upon the release rate, pool size, flame height, and fire duration. Gases, flame, heat, and smoke resulting from the fire cause severe consequences to personnel, equipment, and structure.

A detailed analysis is required to identify the fire scenarios in the offshore platforms. Sufficient empirical equations should be used to calculate the heat transferred from different types of fires. Fires range in size and consequence from small, easily controllable fires causing minor damage to large fires creating

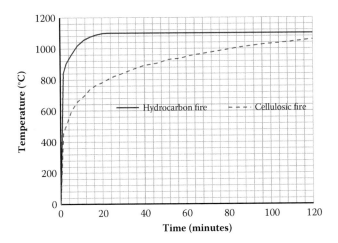

FIGURE 1.13   Growth of Hydrocarbon fire.

a significant loss. The behavior of a structure under fire is profoundly affected by the fire growth rate and the duration of the fire. The duration of the fire depends upon the availability of the fuel, location of the fire, and the environmental conditions. All structural materials exposed to fire are susceptible to failure, and the failure of a structural element occurs when it is unable to resist the load. The time taken for the failure of the structure depends upon the type and amount of heat flux and the nature of exposure. In this context, the hydrocarbon fire poses a severe threat to the structure by affecting the mechanical properties of the material due to the rapid growth of temperature in a short duration. Material properties of steel, such as yield strength, thermal conductivity, specific heat, and thermal strain, are profoundly affected beyond 400°C. The rise in temperature of hydrocarbon fire and that of standard cellulosic fire are compared in Figure 1.13. The temperature under hydrocarbon fire increases to 1100°C within 20 minutes from the onset of a fire, whereas the temperature under cellulosic fire is only about 800°C.

Hydrocarbon fire is one of the most severe accidental loads on an offshore platform, as it may result in the complete collapse of the platform. In the recent past, the fire-resistance design of topside structures of offshore platforms is drawing more attention. The nature of the fire load, performance requirements, and the structure's response under hydrocarbon fire become vital input for the fire-resistant design of the topside. A few of the major accidents that occurred in the past, namely, Piper Alpha (1988), Petrobras (2001), Mumbai High North (2005), Ekofisk (1989), and Deepwater Horizon (2010), reinforce the importance of safety and the structural integrity of the offshore platforms against hydrocarbon fire (Manco et al., 2013; Rivera et al., 2014; Guede, 2017; Hallam et al., 1977). The platform decks should be critically analyzed to assess their behavior under hydrocarbon fire. The structural response of a deck of offshore platforms under elevated temperature is usually evaluated by predicting the survival time of the structure before the occurrence of any form of failure using the standard fire time-temperature curve. The offshore topsides are constructed using different grades of steel. The stress-strain characteristics of steel at elevated temperatures vary significantly for different grades of steel. An increase in temperature develops thermal strains in the material, even in the absence of mechanical loading. So, the structural elements experience thermal strain without an increase in the internal stresses under higher temperatures. At temperatures above 450°C, creep strain also develops in the material. With the rise in temperature, Young's modulus, stiffness, and yield strength of structural steel decreases with or without the development of mechanical strains. The principal mechanism that causes a reduction in the strength and stability of the structure during a fire is the release of potential energy. Through energy absorption, the denser internal structure of steel leads to a phase transformation at 730°C. Thermal conductivity, specific heat, thermal strain, and thermal expansion varies with the increase in temperature.

1.2.5 IMPACT LOADS

Because of the considerable distance from the shore, offshore platforms commissioned in ultra-deep waters require servicing from bigger supply boats and vessels. Hence, column members of the offshore platforms are highly prone to impact loads that may arise from a ship-platform collision. It generates high-impact energy capable of causing both local and global deformations on the column members. With the increase in the number of oil production and exploration platforms, the risk of ship collision and ice impact has also substantially grown in the recent past. Ship-platform collision is a dynamic process that involves several factors, namely, type of collision, the contact time of the collision, energy absorption, and dissipation. Though the collisions may not result in the loss of lives or injuries, economic consequences are significant. The compliance of floating rigs may increase the risk involved in impacts, because of very little or even no redundancy in the structure. Besides, the post-collapse strength of the main structural components of such platforms would be deficient. A relatively small dent of very little thickness will be sufficient to remove the entire design safety factor from the structural member (Harding et al., 1983). A detailed analysis of an example compliant structure under impact loads is discussed in Chapter 4.

1.3 COMPLIANT AND FLOATING PLATFORMS

Compliancy refers to flexibility. It is a general design concept that structures resist loads by the material strength. Various design procedures circumscribe around the material strength and exploit them to achieve the maximum outcome without any significant deformation. Further, a higher order of static indeterminacy is considered as a reserve strength, which is generally derived from the geometric shape and arrangement of members in any structural form. The above concepts of structural design are modified in the case of the design of offshore compliant structures (Nagavinothini and Chandrasekaran, 2019; Chandrasekaran and Nagavinothini, 2018a,b; 2017).

The geometric form of the compliant structures is chosen to allow large deformation, but only as a rigid-body motion. It is essential to note that deformation in structural members' engineering materials is not in the plastic stage. On the contrary, conventional design procedures restrict the deformation of members to a maximum permissible limit. So, the design of offshore compliant structures differs from other traditional structures by one vital factor: form-dominant design. An alternate example of form-dominant design is the shell structure, which is one of the most successful and yet very innovative geometric forms. By enabling large displacements of the structure as a whole, termed as rigid-body motion, compliant offshore platforms resist lateral loads by relative displacement between waves and the structure.

With the improved geophysical exploration, oil and gas deposits in the sea can now be detected to a depth of 12 km. Thus, many new oil deposits have been discovered recently, and 481 larger fields were found between 2007 and 2012 in deep and ultra-deep waters. It is highly essential to note that the newly discovered offshore fields are comparatively larger than recently discovered onshore fields. A variety of offshore structures have been developed to facilitate oil and gas exploration and production from shallow water to ultra-deep-water depths. All offshore platforms consist of two significant components, topsides and substructures. The topsides define the functions to be carried out, supporting structure, and its foundations define the type of platform. Different types of structural geometry emerged to make the platform adaptive at different water depths and environmental conditions. Each structural form was developed mainly by structural innovativeness and suitability in adverse weather conditions. Compliant platforms can be grouped as Guyed Towers, Articulated Towers (AT), Tension Leg Platforms (TLPs), and SPARs. Floating offshore platforms can be further grouped as Semisubmersibles, Drillships, and Floating Production Storage Offloading (FPSO) platforms (Kahla, 1994). Figure 1.14 shows schematic views of compliant and floating platforms.

The design and development of floating offshore structures started in the 1960s due to their better advantages over fixed and compliant structures. They possess a moderately submerged hull, attracting lesser water-plane area. It reduces the effect of the variable submergence when waves pass the structure under motion. It makes them less sensitive to even harsh environmental loadings. Floating offshore platforms are unique in construction, unlike land-based structures. They are designed and constructed to withstand severe dynamic environmental loads, based on the strength of the material and the geometric form.

1.3.1 TENSION LEG PLATFORM (TLP)

Tension leg platforms (TLPs) consist of vertical columns and pontoons, which are position-restrained by tethers. Columns and pontoons are designed as tubular members to enhance buoyancy. Taut-moored tethers balance the excess buoyancy. The size of the platform is chosen to be very large to create excessive buoyancy,

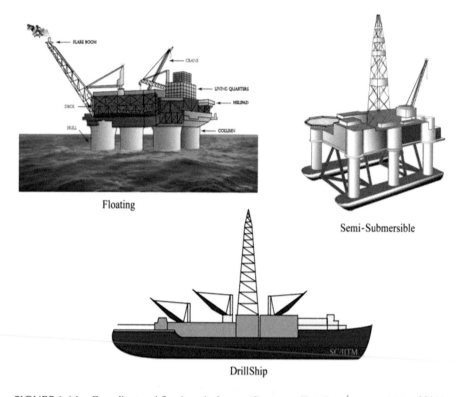

Floating

Semi-Submersible

DrillShip

FIGURE 1.14 Compliant and floating platforms. (Courtesy: Chandrasekaran and Jain, 2010).

which can exceed its self-weight. It enables easy towing of the platform under free-floating condition (Jain and Chandrasekaran, 2004). At the exploration site, the TLP is position-restrained by taut-moored tethers. Excessive buoyancy is transferred to the tethers as initial pre-tension, and hence the name "tension-leg" is tagged to these platforms. TLPs are vertically moored compliant structures with a restrained heave, roll, and pitch motions in a vertical plane. They remain very flexible in the horizontal plane by allowing significant surge, sway, and yaw motions. It makes TLP a hybrid structure. A particular group of motions is very flexible, having a long period, while the other group is rigid, having a short period. It is the novelty of TLP that the platform possesses two distinct groups of frequencies. Figure 1.15 shows the schematic view of the Neptune TLP.

Under the action of wave loads, the platform will move along the direction of waves, which is termed an offset. It also results in motion in the vertical plane, called a set-down. There is a strong correlation between surge and heave motion, which is one of the main reasons for its stability. Under the installed condition, axial tension in the tethers varies continuously, causing a dynamic tension. Hence, the failure of tethers under fatigue is one of the critical viewpoints in the design perspective (Chandrasekaran and Merin, 2016; Chandrasekaran and Visruth, 2013). TLP is capable of operating at water depths in the range of 450 to 2100 m. TLP withholds many operational advantages of fixed platforms but can sustain lighter payloads, unlike fixed platforms.

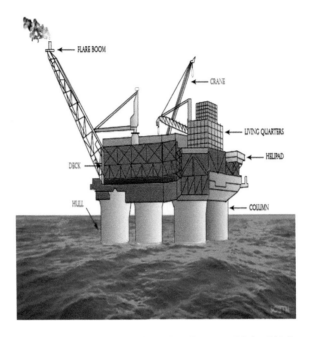

FIGURE 1.15 Neptune TLP. (Courtesy: Chandrasekaran and Jain, 2016).

1.3.2 SINGLE POINT ANCHOR RESERVOIR (SPAR)

A SPAR platform consists of a deep-drafted floating caisson, which is a cylindrical hollow structure similar to a huge buoy (Moo, 2013). The deep draft of the structure provides favorable motion characteristics compared to that of other compliant platforms (Liapis et al., 2010). The cylindrical hull is compartmentalized into different sections. The bottom part, located near the keel, provides buoyancy during transportation and holds ballast materials upon installation. The middle section is used for oil storage, while the upper section provides buoyancy to the structure during operation. A SPAR is position-restrained by a catenary mooring system, which is similar to that of a semi-submersible. But, a SPAR remains stable even when the mooring lines are disconnected, due to deep-draft geometry (Hang et al., 2003; Islam et al., 2017). Based on the configurations of the hull, a SPAR can be classified as a classic, truss, and cell SPAR (Tim et al., 2005). A classic SPAR consists of a large, cylindrical hull with a ballast at the bottom. A truss SPAR consists of a truss-system, which connects the shorter cylinder (called a hard tank) to the soft tank and houses the ballast. A cell SPAR has a large, central cylindrical hull surrounded by small cylinders. SPAR platforms have a low pitch and heave motion compared to other platforms, which makes them safe and operable up to a water depth of about 3000 m. Figure 1.16 shows the schematic view of a SPAR platform.

FIGURE 1.16   SPAR platform. (Courtesy: Chandrasekaran and Jain, 2016).

1.3.3 Semi-Submersibles

Floating offshore platforms are position-restrained generally using the Dynamic Positioning System (DPS), which is referred to as the active restraining system. Semi-submersible platforms are classified as floating facilities, which are position-restrained by mooring lines (Nguyen, 2010). Alternatively, Dynamic Positioning Systems (DPS) are also deployed for this purpose. Semi-submersibles comprise stationery floating hulls supported by pontoons that are tethered to the seabed with mooring lines. These mooring lines can be wire rope, chains, polyester, or any combination of these. Unlike a SPAR, semi-submersibles are not typically designed to accommodate dry-trees. They host one or more subsea systems, which may be deployed to one or more oil fields. Semi-submersibles usually export oil and gas via export risers that form the pipeline systems (Jensen et al., 2010; Niedzwecki and Liagre, 2003). The first FPS was a converted semi-submersible drilling rig, installed on Hamilton Oil Co. Ltd.'s Argyll field offshore the United Kingdom in June 1975. One of the most in-depth semi-submersible production units is Anadarko's Independence Hub, commissioned in 2007, in Gulf of Mexico at a water depth of 2376 m. More details on the geometric forms and response analyses can be seen in further chapters. Figure 1.17 shows the schematic view of the Blind Faith semi-submersible.

1.3.4 Drillships

Drillships are large sea-going vessels, whose hull forms are modified to house drilling and production equipment. They are deployed primarily for exploratory drilling in deep waters under rough sea conditions. A drillship consists of a drilling platform and a

FIGURE 1.17   Blind Faith semi-submersible. (Courtesy: Chandrasekaran and Jain, 2016).

derrick located in the middle of its deck. It also includes a moon-pool, which extends right through the hull of the ship. A moon-pool is primarily an opening through the hull, which facilitates the use of drilling operations. Large hull forms possess excellent hydrodynamic stability but are unstable under combined wave and wind forces. Dynamic positioning systems and moorings are used for station keeping during the drilling operation. Drillships have a large payload capacity and are subjected to a longer period of downtime under wind and wave action. Accurate fatigue assessment of ship structures is an essential part of the structural integrity. Because of accumulated fatigue damage, cracks may occur earlier than expected in critical locations. In many practical cases, the data set needed for accurate fatigue damage estimation can be obtained by either numerical simulation or direct measurement. But, in practice, the

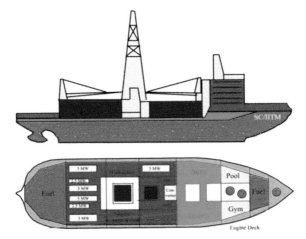

FIGURE 1.18   Drillship. (Courtesy: Chandrasekaran and Jain, 2016).

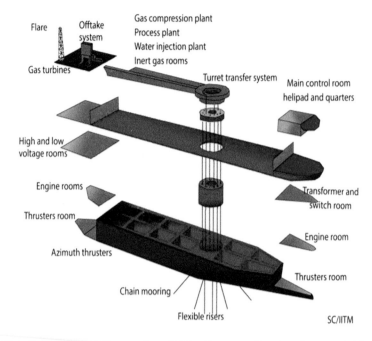

FIGURE 1.19   Processing system of an FPSO. (Courtesy: Chandrasekaran and Jain, 2016).

data set mentioned above is often limited and not large enough for an accurate direct fatigue estimate. Because of their excellent mobility and ready accessibility to exploratory oil fields, Drillships are preferred over semi-submersibles and other permanent complaint platforms. Figure 1.18 shows the schematic view of a Drillship.

1.3.5 Floating Production Storage and Offloading (FPSO) Platform

An FPSO consists of large, barge-shaped structures that are position restrained by turret-moored systems. The turret minimizes the forces on the structure during extreme weather conditions. FPSOs are either newly constructed tankers or custom-made hull forms of ships, useful for production and storage of hydrocarbons. Later, hydrocarbons that are produced will be transported by other vessels to the ports. Flexible risers are used in FPSOs to sustain the significant motion of the platform (How et al., 2009; Lie and Kaasen, 2006; Sparks, 2007). Floating Production Systems (FPS) and Floating, Storage, Offloading (FPO) systems are the design variants of FPSO (Bai and Bai, 2005). These structures are relatively insensitive to water depth and the capacity to disconnect from the moored location, as and when required (Konovessis et al., 2014). It makes them suitable for iceberg prone areas. However, these structures have a low deck capacity and little oil storage capability. Figure 1.19 shows a schematic view of the processing system of an FPSO.

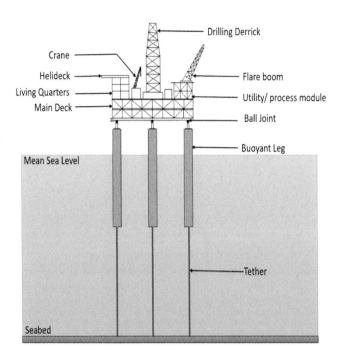

FIGURE 1.20 Triceratops.

1.3.6 TRICERATOPS

The need for an innovative and adaptable structural form paved the way for the development of new-generation offshore platforms (Chandrasekaran and Nagavinothini, 2020; Chandrasekaran, 2019a,b; Chandrasekaran and Madhuri, 2012a,b; Chandrasekaran et al., 2010a). Triceratops is one of the recently conceived new-generation offshore platforms. It consists of a deck and three buoyant legs, which are position-restrained by a set of taut-moored tethers, as shown in Figure 1.20. The most-innovative component that makes Triceratops different from other platforms is the ball joint (Chandrasekaran and Jamshed, 2017; Chandrasekaran and Kiran, 2018; Chandrasekaran and Senger, 2017). Ball joints are used to connect the deck and buoyant legs (Chandrasekaran and Madhuri, 2013;). They restrain the transfer of rotational motion and allow only translational motion between the deck and buoyant legs. Thus, under the action of wave loads on buoyant legs, rotational motion such as roll, pitch, and yaw will not be transferred to the deck (Chandrasekaran, 2017; Chandrasekaran et al., 2013). It, therefore, ensures a convenient workspace for the crew on top of the deck (Chandrasekaran and Madhuri, 2012). These distinct motion characteristics provide uniqueness to Triceratops. It is stiff in the vertical plane and compliant in the horizontal plane, similar to that of TLPs (Chandrasekaran and Nagavinothini, 2017; 2018; Chandrasekaran and Jamshed, 2015; Chandrasekaran et al., 2015a). Buoyant legs of Triceratops resemble the hull of the SPAR platform. Thus, Triceratops derives

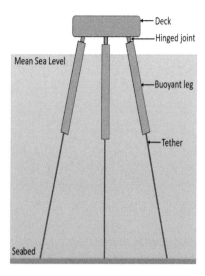

FIGURE 1.21   Buoyant Leg Storage Regasification platform.

advantages from both TLP and SPAR (Chandrasekaran and Vishruth, 2013). Besides, Triceratops attracts lesser forces due to its reduced water-plane area.

1.3.7 Storage and Regasification Platform

To overcome the limitations of the storage capacity of a Floating Storage Regasification platform (FSRU), the Buoyant Leg Storage and Regasification platform (BLSRP) was conceived by researchers (Chandarsekaran and Gaurav, 2017; Chandrasekaran and Kiran, 2018; Chandrasekaran and Jamshed, 2017; Chandrasekaran and Thailammai, 2018; Chandrasekaran and Lognath, 2017). BLSRP consists of a circular deck and six buoyant legs positioned at an inclination of 20° to the deck, as shown in Figure 1.21. The deck consists of regasification equipment, storage tanks, and seawater pumps. Buoyant legs are position-restrained by taut moored tethers, similar to that of a TLP (Chandrasekaran and Lognath, 2017; Chandrasekaran and Kiran, 2017; Chandrasekaran et al., 2015b). The platform remains stiff in the vertical plane and compliant in the horizontal plane. Buoyant legs are connected to the deck by hinged joints, which restrain the transfer of rotational motion and allow only translational motion. Thus, this platform has combined advantages of the TLP, SPAR, and Triceratops (Chandrasekaran and Lognath, 2015; Chandrasekaran and Madhiuri, 2015). The symmetrical geometric form also makes the platform insensitive to the direction of wave load action.

1.4 RISERS AND MOORINGS

Floating offshore structures are dominating in deep and ultra-deep waters in terms of exploration and production of oil and gas, because of their numerous advantages.

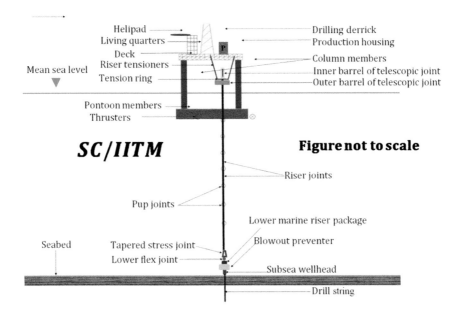

FIGURE 1.22   Pre-tensioned riser connected to a semi-submersible.

During the exploration and production of oil and gas, the vital mechanical component of an offshore platform is the riser. A riser is a closed conduit where one end is attached to the platform with various other mechanical elements and the other end is attached to the sub-sea wellhead; the wellhead rests on the seabed with a permanent and rigid connection (Jeans et al., 2002). Through the annular spacing of the riser, hydrocarbons will be drawn from seabed upwards; hence, they are called "risers," as the flow is against gravity.

A riser is a critical component, which is under the continuous action of wave forces (Mazzilli and Sanches, 2011). Failure of a riser can cause an oil-spill and result in a potential threat to the marine eco-system. It is in addition to the structural failure of the riser. Some vital components of a tensioned riser, coupled to an offshore platform, are a production or drilling riser (as the function may be); riser joints (bare joints or riser joints with buoyancy modules, as the case may be); buoyancy modules; pup joints, which connect each riser joint; a subsea wellhead; a Blow out preventer (BOP), a lower marine riser package (LMRP); a tapered stress joint (TSJ); a lower flex joint (LFJ); a tension ring; a telescopic joint, which connects the outer and inner barrel; hydro-pneumatic direct-acting riser tensioner (HPDART); and an upper flex joint (UFJ). Figure 1.22 shows a few of these components on a riser connected to a semi-submersible. Figure 1.23 shows the elements of a steel catenary riser attached to a semi-submersible.

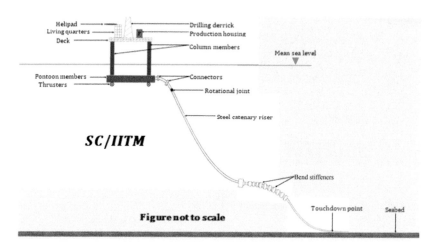

FIGURE 1.23   Catenary riser connected to a semi-submersible.

1.4.1 Classification of the Riser

Based on the purpose and application, risers are classified as drilling riser, production riser, injection riser, workover riser, and export/import risers. Drilling risers
are the primary conduits, deployed for drilling the oil wells. Marine drilling risers
are conduits of a large diameter, consisting of the main tube and external auxiliary
lines. Pressure in the main pipe is kept low so that the pipe can receive input from
the auxiliary lines, which are maintained at a higher pressure. Drilling risers are
connected to the subsea blowout preventer securely. As marine risers are suspended
during installation, they float at greater water depths. Hence, they are controlled
using top-tensioners to ensure the stability of the platform during drilling.
Alternatively, they are also attached to buoyant tubes to reduce the top tension.

Production risers are useful in transporting oil (or gas) continuously from the well
to the top side of the platform. These risers are also larger in diameter and slender in
cross-section. To avoid bending along their length, bend-stiffeners are provided (see
Figure 1.17 for details). Hybrid risers also exist, which is useful for both exploration
and production. The remaining types of risers are deployed for a specific purpose and
can be considered as an ancillary riser network. Risers are also grouped based on the
geometry: catenary riser, flexible riser, and top-tensioned riser. Waves, encountering
the riser, would cause massive displacement. As risers are very long compared to
their cross-sectional dimensions, they need to resist extensive bending. An initial
design was to introduce axial tension of high magnitude along the length to withstand
the lateral forces. Studies showed that such a design would develop high membrane
stresses on the walls of the risers, leading to cracking.

Further, joints will be more vulnerable to failure. To overcome this specific
problem, long, catenary shaped risers are deployed. In a catenary riser, heave motion
compensates for the loss of tension (Gao et al., 2011; Gouveia et al., 2015). Flexible

risers consist of a series of concentric tubes, offering high resistance to bending and axial stresses. They are circumferentially reinforced with armors to enhance their bending capacity. They are quite popular at lesser water depth due to their convenient transport and laying operations.

1.4.2 Functionally Graded Riser Material

Functionally graded material (FGM) is a recent innovation in riser application, though it is widely used in aerospace and mechanical components. Risers transport hydrocarbon and gas at very high temperatures and pressures. They are under a complex failure condition, both structurally and functionally. Structural failure may result from excessive hoop stress developed inside the conduit at high temperatures and pressures. Functional failure refers to material degradation, resulting in corrosion caused by the hydrocarbons. Chlorides, sulfide, and carbon-dioxide gases induce severe corrosion to both the internal and external layers of the riser. Hence, risers should be both strong and corrosion-resistant. Unfortunately, a single material cannot offer the best of both properties. Recent research studies attempted to use functionally raded carbon-manganese steel and X-52 stainless steel to form the FGM. Under the corrosive marine environment, a corrosion-resistant alloy of duplex stainless steel, which is functionally graded with carbon-manganese steel and the inner layer, is investigated by the researchers (Chandrasekaran et al., 2019; 2020).

1.4.3 Mooring Configurations

Mooring configuration should be symmetric to the platform's major axis to avoid rotational motion under wave action. Anchoring is different from a mooring. While the former refers to the operation of holding an object about a fixed point, the latter refers to connecting the moving object to the fixed point, rigidly. Anchoring commonly refers to the holding of any object, but mooring refers to connecting a floating body. In the case of slack-mooring, the length of the mooring is about three times more than the water depth. Anchor points of the mooring should not be widely spread as it may affect the free-access of service vessels to the offshore platform. Simultaneously, the layout of a spread mooring is too close and will not counteract the lateral loads caused by waves (Gottlieb and Yim, 1992). Therefore, the mooring layout plays an essential role in the response behavior of the complaint platform. A mooring system comprises mooring lines, anchors, connectors, clump weights, and other accessories required for anchoring the mooring line to the permanent anchor in the seabed. In some instances, large floating buoys, which are permanently anchored to the seabed, can also be used as temporary moorings. Figure 1.24 shows the anatomy of a Turret-mooring system.

Unlike fixed offshore platforms, both compliant and floating platforms are position-restrained using mooring lines. One end of the mooring line is attached to the platform, at a point called the fairlead, while the other end is anchored to the seabed, called the touchdown point. Moorings will be subjected to the uplift force of larger amplitude at the touchdown point. The design of the mooring system can influence the motion characteristics of floating platforms significantly. Damping

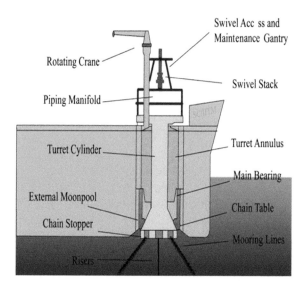

FIGURE 1.24 Anatomy of Turret mooring. (Courtesy: Chandrasekaran and Jain, 2016).

offered by a spread-catenary mooring, with and without a buoy, does not change slow-drift motion (Hassan et al., 2009; Jang et al., 2019). In the case of compliant offshore platforms whose decks are partially isolated, deck response remains independent of the tension variations (Nagavinothini and Chandrasekaran, 2019; 2020). Different sag-to-span ratios and the inclined angle of the mooring lines, under various current velocities also significantly affect the motion response of the floater and the tension in the mooring lines (Wang et al., 2018).

On the other hand, deploying buoys in the mooring system reduces stress in the mooring lines and is found to be effective in deep-water deployment (Xu et al., 2018). Buoys, if provided at various sections along the mooring line, increase the fatigue life of the steel mooring lines (Yan et al., 2018). The influence of the mooring lines on the response behavior of a semi-submersible platform is discussed in detail in the forthcoming chapters. Tension variation in the mooring can result in fatigue failure and needs to be assessed carefully.

2 Semi-Submersibles

2.1 SEMI-SUBMERSIBLES: A PRIMER

Floating offshore platforms are position-restrained generally using the Dynamic Positioning System (DPS), which is referred to as the active restraining system in the literature (Chandrasekaran and Nagavinothini, 2020; Chandrasekaran, 2019a,b; Chandrasekaran and Gaurav, 2017; Chandrasekaran, 2017; Chandrasekaran and Jain, 2016). Alternatively, mooring systems are also deployed to hold down the platform in the desired location while in operation. They are known as the passive restraining system; a combination of both is also not uncommon (Chandrasekaran 2015a,b; 2016a,b; Chandrasekaran, 2014; Chandrasekaran and Bhattacharyya, 2012). As seen in the earlier chapter, semi-submersibles command many advantages as they possess a sizeable water-plane area compared to their competitors. This enables the platform to remain insensitive to the harsh ocean environment. Therefore, the body motion of the semi-submersibles is relatively smaller than that of other small-sized, floating platforms.

Semi-submersibles also possess superior construction and installation features and the convenience of being able to be towed to different locations (Chandrasekaran and Uddin, 2020; Odijie, 2017; Zhu and Ou, 2011). Since the 1980s, there has been extensive research done on the improvement of the hull-design of the semi-submersibles for obtaining a better response under combined environmental loadings. Semi-submersibles are specialized marine vessels, used for various operations as exploratory drilling rigs, safety vessels, oil production platforms, and crane vessels. Unlike fixed platforms that derive their strength from member and material strength, semi-submersibles derive their strength from being highly compliant in the horizontal plane. For any floating structure, all the six-degrees-of-freedom remain active under the action of waves. There are translational degrees-of-freedom along the X, Y, and Z axes, which correspond respectively, to surge, sway, and heave, respectively; rotation about each of these axes are termed roll, pitch, and yaw, respectively. Out of these six degrees-of-freedom, natural periods in surge, sway and yaw motion of a semi-submersible are distinctly different and larger than heave, roll and pitch. Semi-submersibles need to be highly versatile. In the horizontal plane, they need to be flexible, while in the vertical plane, they need to behave stiffly due to their large buoyancy.

2.1.1 GEOMETRIC CONFIGURATION

It is obvious to correlate the fact that semi-submersibles should possess a large geometric shape, thus causing a high displaced volume to induce the desired buoyancy. Semi-submersibles have proportional dimensions along their length and breadth. They are unlike ships, which have lengths significantly larger than their breadth, classifying them as stream-lined bodies. Semi-submersibles are towed to the site and ballasted and moored (anchored) to the seabed while in operation. They have large vertical columns typically connected to two or more air-filled steel floats, called pontoons, that are held in position by massive anchors. Because the pontoons are usually submerged a few feet below the water surface, they are called semi-submersibles. Columns support the deck and equipment on the top side. Semi-submersibles are compliant-type drilling structures, one of the oldest offshore exploratory rigs used in the deep sea. Most of the semi-submersible platforms are rectangular in plan with the aspect ratio (L/B) of about 1.25–1.40. A large plane area is desirable to perform the intended functions of the platform. Further, a multi-tier deck system is not preferred, unlike other compliant platforms, as this may induce more rotational response in the vertical plane. The basic geometric design of a semi-submersible is typically rectangular with lesser height. The vertical center of gravity lies within the operational plane; cranes and drilling derricks, placed on the deck, are exemptions. It improves operational stability even under harsh weather. It also enhances maneuverability, making the platform more mobile. Mobile Offshore Drilling Units (MODUs), which are one of the unique forms of semi-submersibles, are commonly used for exploratory drilling. Figures 2.1, 2.2, and 2.3 show different semi-submersibles commissioned at various locations. Table 2.1 provides a summary of semi-submersibles installed at multiple locations around the world.

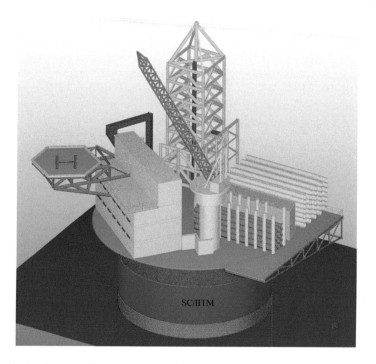

FIGURE 2.1   Sevan Driller semi-submersible.

FIGURE 2.2   West Alpha semi-submersible drilling rig.

2.1.2 FUNCTIONAL REQUIREMENTS

In general, it is well-understood that compliant and floating offshore platforms undergo large displacements due to high flexibility in the horizontal plane. It includes significant surge, sway, and yaw motions, which are undesirable to perform drilling operations. Therefore, the geometric forms of semi-submersibles are the modified forms of ships to suit the functional requirements, like drilling, production, and storage. Unlike ships, the operational deck of a typical semi-submersible (see Figures 2.1, 2.2, and 2.3) is densely packed with plants and types of equipment. The drilling derrick, pipelines, and cable trays used to carry the electric accessories contribute to the congestive layout of the top side. In addition to the living quarters and other civil accessories that are required to be housed, about 50–75 personnel on board perform various operations around the clock. Considering the above factors, it is imperative to control the rotational motion of semi-submersibles, particularly in translational motion, in general. It is necessary to improve the operational comfort of the personnel on board and the safety of risers and moorings that are connected to the semi-submersibles (Phifer et al., 1994; Salama et al., 1998; 2002). It is also evident from the literature that large mono-hull ships, which were used for drilling earlier, were gradually replaced (or converted) with semi-submersibles in the early 1980s.

A semi-submersible rig floats on the water surface when towed from one drilling location to another. Upon reaching the target, specific compartments of the platform are flooded to submerge the lower part of the rig to the seafloor. The lower part of the rig rests on the seafloor to enable drilling operations. With the base of the rig in contact with the seabed, it has excellent resistance to lateral forces. It is interesting to note that semi-submersibles possess high stability due to deep-draft, achieved during

FIGURE 2.3 Saipem Crane vessel.

such submergence (Lee et al., 2005; Sunil and Mukhopadhyay, 1995). Because of the large shape and size, they also remain positive-buoyant while in operation. This improves the operational safety of the platform, even in harsh sea conditions. Wave loads have little effect on the structural motion of the semi-submersibles. They possess excellent stability during drilling as the top side weight is balanced by the equivalent water-plane area of the submerged part. Compared to the drillships, which float while drilling, semi-submersibles are position-restrained to the seafloor during operation. Some semi-submersibles are also equipped with propellers, to enable smooth, self-sailing capabilities during towing operations. They have inbuilt power units, which can be used to propel them from one site to another.

With the pontoons and columns submerged at a deep-draft, a semi-submersible is less affected by waves, in comparison to that of regular ships. As semi-submersibles do not rest on the seafloor, they do not derive strength and stability from the anchors, but obtain it instead through large buoyancy and geometric shape only. It is, therefore, essential to note that semi-submersibles are carefully trimmed to maintain stability. Semi-submersibles can transform from a deep to a shallow draft by de-ballasting and making them like surface ships. Please recollect that de-ballasting refers to removing ballast water from the hull, which is pumped out to sea.

2.1.3 Commissioned Semi-Submersibles

The operational depth of semi-submersibles varies from 90 to 1000 m. High mobility with a high transit speed of about 10 knots makes these platforms very versatile. As their structural form is similar to that of ships or any other sizeable floating vessel, they remain stable and show minimal response under waves. Their sea-keeping characteristics resemble ships that are designed to withstand forces that arise during critical sea states. Limiting dry-dock facilities to repair and difficulties in handling the mooring systems make their choice as a drilling and production platform more exclusive. The first semi-submersible, the Blue Water Rig, was commissioned in 1962 by the Shell Oil Company in the Gulf of Mexico. The geometric design was not successful as it was not sufficiently buoyant. Successively, the plan was altered as Ocean Driller and commissioned in 1963. Table 2.2 shows the summary of essential semi-submersibles commissioned worldwide. Places of operation are shown where they were initially commissioned (Chandrasekaran and Jain, 2016).

2.2 EXAMPLE CASE STUDY 1: NUMERICAL RESPONSE ANALYSIS OF A SEMI-SUBMERSIBLE

Semi-submersibles are position-restrained using mooring lines. Because of the extensive motion of the platforms, moorings undergo dynamic tension variations. Besides the damping offered by the mooring lines, other factors, namely, mooring pre-tension, stiffness, drag coefficient, wave frequency, amplitude, and current, affect the platform (Webster, 1995; Kurian et al., 2010; Lesage and Garthshore, 1987). In the case of moorings with high initial pre-tension, damping is reduced as

TABLE 2.1
Selected semi-submersibles commissioned around the world

Name of the vessel	Oil field location	Region	Year of operation	Size	Displacement (ton)	Water depth (m)	Drilling depth (m)	Station keeping
ARGYLL, FDU	Argyll	North Sea, UK	1975	-----	34000	150	----	DPS
BUCHAN-A	Buchan	North Sea, UK	1981	-------	18995	160	---	DPS
P-09	Corvina	North Sea, UK	1983	-----	22896	230	-----	DPS
P-15	Pirmana	Brazil	1983	-----	36100	243	-----	DPS
P-12	Lingnado	Brazil	1984	-------	22243	100	---	DPS
P-21	Badejo Salema	Brazil	1984	-----	10765	112	-----	DPS
WEST ALPA	--	North Sea	1986	98 x 76m	57563	3000	7000	8 anchor
P-22	Morela	Brazil	1986	-----	17440	114	-----	DPS
P-07	Bicudo	Brazil	1988	--------	20493	207	-----	DPS
P-20	Marlim	Brazil	1992	-----	25983	625	-----	DPS
P-08	Marimba	Brazil	1993	-----	20990	423	---	DPS
P-13	Bijupura Salena	Brazil	1993	--------	22243	625	-----	DPS
P-14	Caravela	Brazil	1993	-------	21616	195	-----	DPS
P-18	Marlim	Brazil	1994	-----	33400	910	---	DPS
P-25	Albacora	Brazil	1996	-----	25983	252	-----	DPS
P-27	Vaedor	Brazil	1996	-------	41659	533	-----	DPS
P-19	Marlim	Brazil	1997	-----	33400	770	-----	DPS
P-26	Marlim	Brazil	2000	--------	27656	515	-----	DPS
P-36	Rancador	Brazil-Campos	2000	--------	----	1360	---	DPS
P-51	Marlim Sul	Brazil	2000	-----	80114	1255	---	DPS
SS-11	Coral	Brazil	2003	-------	---	145	---	DPS
P-40	Marlim Sul	Brazil	2004	-------	---	1080	-----	DPS
SEVAN DRILLER	--	Norway	2009	75 m dia	58453	3000	10000	DPS-3
WEST AQUARIUS	--	Panama	2009	117 x 78m	40731	600	10668	DPS-3
P-56	Marlim Sul	Brazil	2010	-------	50000	1700	-----	DPS
HAI YANG SHI YOU-981	----	China	2010	114 x 90m	30670	1500		
WEST CAPRICON	--	Panama	2011	116 x 90m	30147	3048	---	---
SEVAN BRASIL	Petrobras	Brazil	2012	86 x 75m	55580	3000	10000	DPS-3
P-55	Rancador	Brazil	2012	-------	105000	1707	---	DPS
SEVAN LOUSIANA	--		2013	99 m dia	58650	3048	10668	Kongsberg
P-52	Rancador	Brazil	2017	-------	80201	1795	-----	DPS

TABLE 2.2

Semi-submersibles commissioned 1 word: worldwide

S. No	Continent	No. of platforms
1	Europe	16
2	North America	7
3	South America	22
4	Asia	3

the drag coefficient increases; this is due to higher resistance to transverse motion. Hybrid mooring lines of the spread mooring system, consisting of a chain-wire configuration, are also deployed in a few semi-submersibles (Garza Rios et al., 2000; Kim et al., 2019). Researchers confirmed that the platform could face stability issues under the change in initial pre-tension and angle of inclination of the moorings. Following these observations, recently commissioned semi-submersibles are deployed with mooring lines at the lesser angle of inclination. The basic idea is

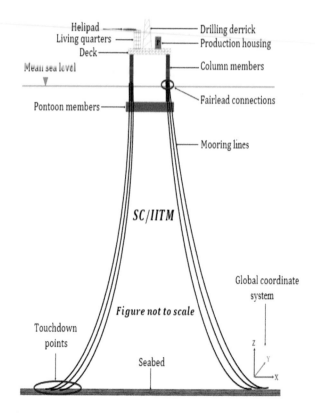

FIGURE 2.4   Semi-submersible without production riser.

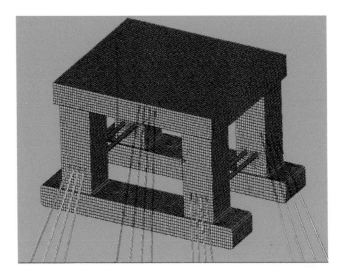

FIGURE 2.5   Numerical model of the semi-submersible.

to achieve the desired dynamic behavior of the platform in the horizontal plane. An increase in the angle of mooring lines also increases the axial tension but can offer better resistance against rotational degrees-of-freedom (Chen et al., 2011). One of the attractive features of the semi-submersibles is their adaptability to water depth. But, water depth is found to be an essential factor affecting stability. The geometric design of the semi-submersibles is, therefore, altered appropriately with the truss-

TABLE 2.3

Description of a semi-submersible

Particulars	Value
Deck cross-section	74.42 × 74.42 × 8.60 m
Column members	17.385 × 17.385 × 21.46 m
Pontoon members	114.07 × 20.12 × 8.54 m
Displacement	48206.8 tons
Water depth	1500 m
Draft	−21 m
Spacing between columns	55 m
Diameter of the brace	1.8 m
Center of gravity below water surface	−9.5 m
Metacentric height	16.03 m
Radius of gyration for Roll	32.4 m
Radius of gyration for Pitch	34.1 m
Radius of gyration for Yaw	34.4 m

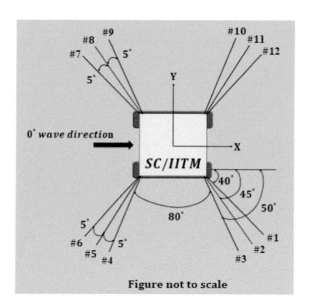

FIGURE 2.6   Layout of the spread mooring system.

pontoon type. Additional heave plates are inducted to counteract the resonating response under extreme waves (Srinivasan et al., 2006).

Recent studies have showed that there is a significant reduction in the fatigue life of the mooring lines when they are taut-moored (Chandrasekaran and Uddin, 2020; Xue et al., 2018). Fatigue life decreases with the initial pre-tension (Huang et al., 2011; Yuan et al., 2011). As semi-submersibles possess large buoyancy, higher initial pre-tension is necessary if taut-moorings are deployed. Therefore, a large number of semi-submersibles in the recent past are deployed with slack-mooring systems. Studies carried out on the response analysis of semi-submersibles with two different moorings, namely, chain-wire-chain and chain-polyester rope-chain, recommended the use of a hybrid mooring system under such a corrosive marine environment (Qiao et al., 2014; Qiao and Ou, 2013). Surge response and tension variation in the mooring lines were found to be significantly higher at the low-frequency excitations. Damping offered by the chain-polyester rope-chain type mooring was found to be higher than that of the chain-wire-chain type. It is due to the large diameter of the polyester ropes, while they are found to be effective in reducing the surge response.

Semi-submersibles are required to be moored to the seabed while in operation. Further, the failure of moorings can lead to a free-surface condition of the platform. It is essential to realize that semi-submersibles cannot function as drilling units under such conditions, because they derive their strength only from deep-draft situations. Therefore, the failure of mooring lines poses a severe threat to the platform's functional safety, not to its structural safety. Studies carried out by Wu et al. (2015) on the fatigue damage of mooring systems showed that the fatigue damage occurs at the top of the bottom chain under taut-moored conditions. But, for a catenary mooring, hot spots of the fatigue failure shift to the fairlead point. Hence, the location of the fatigue damage may alter with the

TABLE 2.4

Configuration of a spread mooring system

Mooring type	Length of the mooring system (m)			Pre-tension (kN)
	Upper chain	Middle wire	Lower chain	
Catenary	600	2000	1600	3000
Taut	600	1300	200	2570

type of mooring configuration, mooring-seabed friction, and other environmental conditions like wave climate. The fatigue life of the mooring under the varying tensions caused by the significant motion of the semi-submersibles can be carried out using various approaches (Yang et al., 2016; Al-Solihat et al., 2016; Chandrasekaran and Uddin, 2020). The S-N curve, T-N curve, and fracture mechanics approaches are conventional and more recent (Chandrasekaran, 2017; 2014; 2019a, b). It is well established from the studies that the fatigue life depends on various factors, namely, initial crack location, its propagation, and depth. Based on the results obtained from the three approaches above, it is seen that the crown section of the mooring chain is more prone to damage in comparison to other weld sections, without considering the stress concentration factor (Chandrasekaran and Nagavinothini, 2018a,b; 2017; 2019a; Chandrasekaran and Bhattacharyya, 2012). Further, the fatigue damage estimated by the T-N curves is more conservative than that of the S-N curve approach.

As seen in the above discussions, the failure of mooring lines should be avoided, but this may be impossible due to the large displacement of the platform. It is also understood that mooring lines fail due to their coupling with the platform motion, which is significantly large. It is also worth noting that semi-submersibles with spread moorings possess improved motion response (DNV, 2005; 2016). To reduce the response amplitude of the platform in the complaint degrees-of-freedom, Yan et al. (2016; 2018) proposed a new configuration of the mooring system, attached with the submerged buoys. They showed that the position and size of the buoy influences reduction in the surge response of the semi-submersible. However, surge motion at a low frequency dominates and increases with the size of the submerged buoys. Hence, an example study

TABLE 2.5

Structural properties a of spread mooring system

Description	Upper chain	Middle wire	Bottom chain
Mass per unit length (kg/m)	163.86	36.41	163.86
Equivalent cross-section (m^2)	0.014	0.014	0.014
Axial stiffness (MN/m)	676.81	833.91	676.81
Equivalent diameter (m)	0.095	0.095	0.095
Longitudinal drag coefficient	0.025	0.025	0.025
Transverse drag coefficient	2.4	1.6	2.4
Added mass coefficient	1.0	1.0	1.0

TABLE 2.6

Environmental conditions for the study

Description	Return period	
	10 years	**100 years**
Significant wave height (m)	11.1	13.3
Peak spectral period (s)	13.6	15.5
Peakedness parameter (Υ)	5	7
Wind speed (m/s)	48.3	55
Current speed (m/s)	1.7	1.97

on the response analysis of semi-submersibles is considered with a spread-mooring system. It will be dealt with in detail in the next section of this chapter.

2.2.1 DESCRIPTION OF THE PLATFORM

The semi-submersible, selected for the present study, is shown in Figure 2.4. The name of the platform and the commissioned location are masked for strategic reasons, but the numerical model reflects the complete details of the platform. The numerical model consists of two horizontal pontoons, partially submerged to maintain the stability of the platform. Four vertical columns support the deck, which, in turn, supports the drilling derrick, production housing, machinery, and other drilling accessories. Four horizontal braces connect the column members to maintain an integral module of the complete unit. The platform is position-restrained with a twelve-point, spread mooring system. The description of the semi-submersible is shown in Table 2.3. Figure 2.5 shows the numerical model.

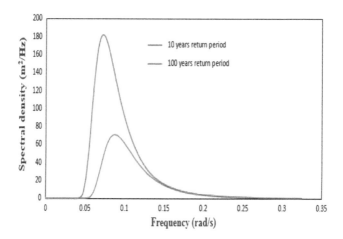

FIGURE 2.7 JONSWAP spectrum.

TABLE 2.7
Free-decay test results under moored conditions

Description	Catenary mooring		Taut mooring	
	Natural period (s)	Damping (%)	Natural period (s)	Damping ξ (%)
Surge	184.58	6.15	148.3	7.12
Heave	21.32	2.47	20.96	1.5
Pitch	25.36	0.93	25.04	2.45

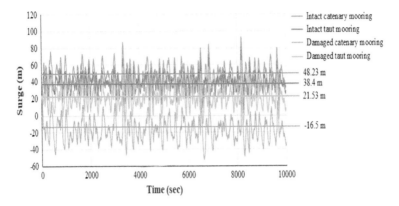

FIGURE 2.8   Surge response of semi-submersible (10-year return period).

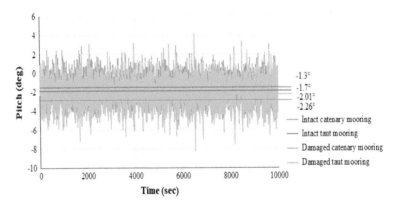

FIGURE 2.9   Pitch response of semi-submersible (10-year return period).

2.2.2 MOORING SYSTEM

In the present study, a semi-submersible is position-restrained with a spread mooring (both catenary and taut) with 12 mooring lines. Figure 2.6 shows the layout configuration of the mooring lines. Mooring lines used in the present study are in the form of chain-wire-chain arrangement. The top and bottom

TABLE 2.8

Semi-submersible motion statistics without mooring failure (10-year return period)

Response	Statistical values	Spread catenary mooring			Spread taut mooring		
		0°	45°	90°	0°	45°	90°
Surge (m)	Minimum	20.70	18.35	22.39	3.02	13.83	22.78
	Maximum	93.00	53.84	22.91	77.43	48.81	23.36
	Mean	48.23	34.75	22.62	38.40	30.68	23.04
	Standard deviation	11.80	5.38	0.08	9.91	4.94	0.08
Sway (m)	Minimum	−5.39	−6.70	−10.08	−6.93	−12.06	−15.49
	Maximum	−4.66	16.09	27.79	−6.22	9.81	20.33
	Mean	−5.05	3.75	7.40	−6.59	−1.01	1.64
	Standard deviation	0.12	3.63	5.45	0.11	3.37	5.07
Heave (m)	Minimum	−13.45	−12.96	−12.63	−12.43	−12.09	−11.68
	Maximum	−6.18	−5.84	−6.94	−5.26	−5.02	−6.04
	Mean	−10.17	−10.17	−10.2	−8.98	−8.93	−8.95
	Standard deviation	0.73	0.77	0.68	0.72	0.74	0.65
Roll (deg)	Minimum	−3.00	−5.00	−7.53	−3.16	−4.67	−7.44
	Maximum	−2.41	−0.03	1.16	−2.43	0.43	1.40
	Mean	−2.70	−2.58	−2.58	−2.79	−2.51	−2.45
	Standard deviation	0.06	0.63	1.11	0.07	0.66	1.14
Pitch (deg)	Minimum	−8.21	5.60	−1.27	8.51	−6.15	−1.3
	Maximum	3.79	2.35	−1.02	4.21	2.78	−1.08
	Mean	−1.30	−1.18	−1.13	−1.7	−1.39	−1.17
	Standard deviation	1.22	0.92	0.02	1.27	0.901	0.02
Yaw (deg)	Minimum	−0.78	−3.22	−0.10	−0.97	−5.97	−0.21
	Maximum	0.51	4.14	0.11	0.65	6.09	0.16
	Mean	−0.04	0.24	0.017	−0.11	0.36	0.01
	Standard deviation	0.16	0.93	0.02	0.25	1.71	0.04

sections comprise a K-4 stud-less chain, while the mid-section is of a spiral strand steel wire. The included angle between the mooring lines within a bundle is 5°, while the angle between each bunch of the mooring lines is 80°, as shown in the figure. Mooring configuration details and the mechanical properties of the moorings are summarized in Tables 2.4 and 2.5, respectively.

2.2.3 Environmental Forces

The Gulf of Mexico (GoM), South China Sea, Persian Gulf, North Sea, etc., are well-known offshore regions where oil and gas exploration and production is carried out. In the present study, the Eastern Gulf of Mexico region (American Petroleum Institute, 2007; 2018) is considered for 0°, 45°, and 90° wave directions;

TABLE 2.9
Semi-submersible motion statistics under postulated failure of mooring (10 Y)

Response	Statistical values	Spread catenary mooring			Spread taut mooring		
		0°	45°	90°	0°	45°	90°
Surge (m)	Minimum	−52.28	−50.08	−65.36	−9.17	−6.82	−8.93
	Maximum	31.83	23	22.41	65.26	36.42	22.92
	Mean	−16.5	−31.21	−46.87	21.53	12.79	2.9
	Standard deviation	13.76	7.47	4.43	11.2	6.55	2.5
Sway (m)	Minimum	−87.15	−76.12	−73.7	−40.26	−35.64	−34.2
	Maximum	−5.22	−4.7	−4.86	−1.46	1.84	3.52
	Mean	−63.47	−56.1	−55.69	−21.41	−17.05	−16.58
	Standard deviation	6.12	6.25	6.69	6.9	5.94	5.7
Heave (m)	Minimum	−13.22	−12.63	−12.22	−12	−11.94	−11.53
	Maximum	−6.19	−5.62	−6.17	−5.28	−5.33	−5.84
	Mean	−9.59	−9.59	−9.62	−8.61	−8.56	−8.61
	Standard deviation	0.79	0.81	0.7	0.75	0.78	0.67
Roll (deg)	Minimum	−2.7	−3.6	−5.98	−3.84	−4.6	−6.71
	Maximum	0.63	1.39	2.6	−0.87	0.62	1.77
	Mean	−1.52	−1.39	−1.39	−2.37	−2.08	−2.02
	Standard deviation	0.22	0.63	1.12	0.4	0.67	1.13
Pitch (deg)	Minimum	−7.24	−5.97	−4.24	−8.22	−6	−2.91
	Maximum	3.62	1.61	−0.7	4.24	2.72	−0.18
	Mean	−2.26	−2.15	−2.09	−2.01	−1.72	−1.48
	Standard deviation	1.29	1.05	0.14	1.31	1.03	0.16
Yaw (deg)	Minimum	0.06	−2.44	0.029	−1.26	−4.63	0.03
	Maximum	3.17	6.65	3.38	4.04	9.52	3.63
	Mean	1.79	2.31	2.11	1.51	2.34	1.91
	Standard deviation	0.39	1.07	0.16	0.73	2.14	0.35

environmental conditions are shown in Table 2.6 (Qiao and Ou, 2013). The Joint North Sea Wave Project (JONSWAP) spectrum is used for characterizing the wave loads, while the API wind spectrum is used for the wind load; the effect of one-hour mean wind speed is considered for the study. The impact of nonlinearly varying current is considered up to a depth of 150 m. Figure 2.7 shows the wave spectral plot used in the present study.

Wave forces acting on the semi-submersible are estimated using the boundary element method (BEM), which is based on the diffraction theory. For irregular waves with slow-drift, the JONSWAP spectrum is used, which includes the imbalanced flow of energy in the waves, generated due to high wind speeds. For a particular frequency, the spectral ordinate is given by the following relationship:

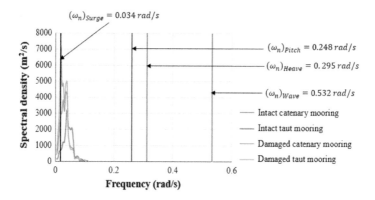

FIGURE 2.10 Surge spectral density plot under 0° wave heading.

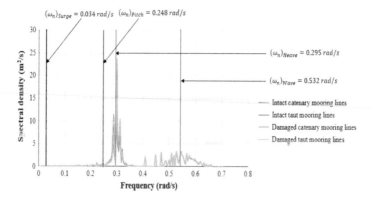

FIGURE 2.11 Heave spectral density plot under 45° wave heading.

$$S\left(\omega\right) = \frac{\alpha g^2 Y^a}{\omega^5} e^{\left(\frac{-5}{4}\left(\frac{\omega_p}{\omega}\right)^4\right)}$$ (2.1)

$$a = e^{\left[\frac{-\left(\omega - \omega_p\right)^2}{2\sigma^2 \omega_p^2}\right]}$$ (2.2)

$$\sigma = \begin{cases} 0.09 \, for \, \omega > \omega_p \\ 0.07 \, for \, \omega \leq \omega_p \end{cases}.$$ (2.3)

where g is the acceleration due to gravity (m²/s), Y is the peak enhancement factor, α is the spectral energy parameter, T_p is the peak wave period (sec), ω is the

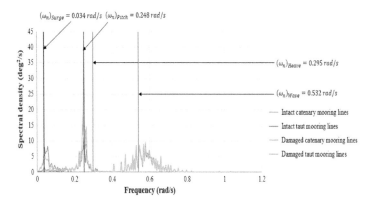

FIGURE 2.12   Pitch spectral density plot under 90° wave heading.

frequency (rad/s), ω_p is the peak frequency (rad/s), H_s is the significant wave height, and σ is the spectral parameter. The spectral energy parameter, used in Eq. (2.1) is given by the following relationship:

$$\alpha = \frac{H_s^{\,2}}{16 \int_0^\infty \frac{g^2 \Upsilon^a}{\omega^5} \times \exp\left[\frac{-5}{4} \times \left(\frac{\omega_p}{\omega}\right)^4\right] d\omega} \tag{2.4}$$

Wind fluctuations cause low-frequency motion on the semi-submersible. The following relationship gives the API wind spectrum, used in the present study with one-hour mean wind speed at a reference height 'Z' (m):

$$S\left(\tilde{f}\right) = \frac{\tilde{f}}{\left(1 + 1.5\tilde{f}\right)^{5/3}} \tag{2.5}$$

$$\tilde{f} = \frac{f}{f_p} \text{ and } f_p = 0.025\left(\frac{\bar{V}_z}{Z}\right) \tag{2.6}$$

$$\bar{V}_z = \bar{V}_{10}\left(\frac{Z}{10}\right)^{0.125} \tag{2.7}$$

where $\tilde{f}$ is the non-dimensional frequency, f' and f_p' are frequencies (Hz), and $\bar{V}_z$ is the mean 1-hour wind speed (m/s). Based on the above equations, the API wind spectral energy density (m²/s) is given by the following relationship:

TABLE 2.10

Mooring tension statistics (10-year return period)

Mooring lines	Mooring tension statistics	Spread catenary mooring			Spread taut mooring		
		0°	45°	90°	0°	45°	90°
# 1 (MN)	Minimum	0	0	0	0	0	0
	Maximum	4.26	4.26	4.27	2.5	2.5	2.5
	Mean	0.00017	0.00017	0.00017	0.0017	0.0018	0.0014
	Standard deviation	0.02	0.027	0.027	0.06	0.06	0.05
# 2 (MN)	Minimum	4.26	4.26	4.23	1.63	1.57	2.14
	Maximum	6.7	6.67	7.08	5.89	6.37	6.03
	Mean	5.57	5.67	5.97	3.11	3.22	3.64
	Standard deviation	0.33	0.27	0.23	0.67	0.61	0.56
# 3 (MN)	Minimum	3.25	3.2	3.11	1.75	1.67	1.33
	Maximum	4.3	4.27	4.3	3.83	3.08	2.69
	Mean	3.63	3.6	3.53	2.19	2.13	2.01
	Standard deviation	0.13	0.12	0.12	0.23	0.17	0.17
# 4 (MN)	Minimum	3.42	3.88	4.05	1.45	1.82	1.93
	Maximum	4.58	4.63	4.89	3.59	3.03	3.57
	Mean	4.12	4.3	4.46	2.23	2.43	2.68
	Standard deviation	0.15	0.1	0.1	0.22	0.15	0.23
# 5 (MN)	Minimum	3.25	3.2	3.12	1.76	1.68	1.33
	Maximum	1.27	4.27	4.3	3.56	3.06	2.73
	Mean	3.63	3.6	3.54	2.17	2.12	2.02
	Standard deviation	0.12	0.12	0.12	0.21	0.17	0.17
# 6 (MN)	Minimum	3.42	3.84	3.95	1.49	1.82	1.87
	Maximum	4.48	4.55	4.79	3.42	2.96	3.52
	Mean	4.05	4.22	4.36	2.21	2.39	2.6
	Standard deviation	0.14	0.098	0.1	0.2	0.14	0.23
# 7 (MN)	Minimum	4.08	3.98	3.68	1.96	1.91	1.39
	Maximum	5.38	4.76	4.57	5.71	3.32	3.01
	Mean	4.55	4.34	4.21	2.89	2.48	2.3
	Standard deviation	0.16	0.1	0.12	0.43	0.16	0.22
# 8 (MN)	Minimum	3.43	3.93	4.16	1.41	1.82	1.98
	Maximum	4.68	4.73	5	3.77	3.1	3.61
	Mean	4.19	4.38	4.57	2.26	2.47	2.76
	Standard deviation	0.17	0.1	0.1	0.24	0.16	0.23
# 9 (MN)	Minimum	4.02	3.91	3.6	1.91	1.9	1.37
	Maximum	5.36	4.71	4.49	5.96	3.31	2.92
	Mean	4.47	4.25	4.12	2.83	2.44	2.25
	Standard deviation	0.17	0.1	0.12	0.44	0.16	0.21
# 10 (MN)	Minimum	0	0	0	0	0	0
	Maximum	4.26	4.26	4.27	2.51	2.5	2.51

(Continued)

TABLE 2.10 Mooring tension statistics (10-year return period) (*Continued*)

Mooring lines	Mooring tension statistics	Spread catenary mooring			Spread taut mooring		
		0°	45°	90°	0°	45°	90°
	Mean	0.00017	0.00017	0.00017	0.0017	0.0018	0.0014
	Standard deviation	0.027	0.027	0.02	0.066	0.067	0.059
# 11 (MN)	Minimum	3.25	3.19	3.11	1.75	1.67	1.33
	Maximum	4.33	4.27	4.3	4.13	3.09	2.66
	Mean	3.65	3.6	3.52	2.22	2.13	2.01
	Standard deviation	0.13	0.12	0.12	0.25	0.18	0.17
# 12 (MN)	Minimum	3.92	3.84	3.53	1.87	1.88	1.35
	Maximum	5.33	4.66	4.41	6.18	3.29	2.84
	Mean	4.38	4.17	4.04	2.78	2.4	2.21
	Standard deviation	0.17	0.1	0.12	0.44	0.15	0.2

$$S(f)\frac{\sigma(z)^2}{f}S(\tilde{f}) \qquad (2.8)$$

2.2.4 RESPONSE UNDER SPREAD MOORING

Numerical analysis is carried out to obtain the dynamic response of the semi-submersible under the spread-mooring layout. A time-domain analysis is carried out to represent the effects of the semi-submersible and mooring system's coupled response at every instant of time. Change in the mooring tension under waves and the platform motion is modeled as below:

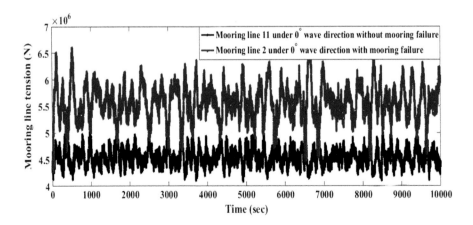

FIGURE 2.13   Tension in catenary mooring lines #11 and #2.

$$\left[M + M_a \right] \ddot{x} + [C] \dot{x} + [K]x = \left\{ F_{static} + F_{WF} + T_M \right\} \qquad (2.9)$$

where [M] is the structural mass matrix (kg), [M_a] is the added mass matrix (kg), [C] is the damping matrix (Ns/m), [K] is the stiffness matrix (N/m), {F_{static}} is the static load (N), {F_{WF}} is the first-order wave loads (N), and {T_M} is the tension in the mooring system (N). Simulations are carried out for 10000 sec with a time step of 0.1 sec. It is interesting to note that numerical simulations need to be carried out for storm durations of 3 hours, which is about 10800 sec. It is necessary to achieve a steady-state of the response history for floating and compliant structures. However, due to the limitations in the software, numerical simulations are limited to 10000 sec in the current study. Simulations at every instant integrate the accelerations obtained in the time domain using a predictor-corrector numerical scheme to get the response history of the semi-submersible. Table 2.7 shows the natural periods of the semi-submersible under moored conditions. It can be seen that even under moored conditions, the platform is highly flexible in the horizontal plane with long periods in the surge and sway motion. Under the rotational motion, it is seen that the platform in more flexible in the horizontal plane, as the yaw motion period is larger than that of the pitch and roll motion. It is also interesting to note that the mooring lines offer a good degree of damping in the compliant degrees-of-freedom.

Figures 2.8 and 2.9 show the Response Amplitude Operator (RAO) for the surge and pitch motions of the semi-submersible, under moored conditions. A comparison is made with the existing studies to validate the present model (Zhai et al., 2011). Surge and pitch RAOs showed a shift in the peak frequency toward the higher side.

TABLE 2.11

Description of a semi-submersible with a submerged buoy

Particulars	Value
Deck cross-section	74.42 × 74.42 × 8.60 m
Column members	17.385 × 17.385 × 21.46 m
Pontoon members	114.07 × 20.12 × 8.54 m
Displacement	48206.8 tons
Water depth	1500 m, 2000 m (2 cases)
Draft	−19 m
Spacing between columns	55 m
Free-board	19.6 m
Diameter of the brace	1.8 m
Center of gravity below water surface	−8.9 m
Metacentric height	16.03 m
Radius of gyration for Roll	32.4 m
Radius of gyration for Pitch	34.1 m
Radius of gyration for Yaw	34.4 m

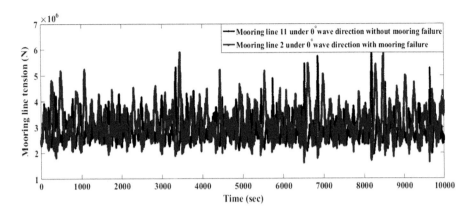

FIGURE 2.14   Tension in the taut-mooring lines #11 and #2.

while the magnitudes are almost agree well with each other. The change in the frequencies is marginal, and hence over-sighted.

A detailed dynamic analysis is carried out under moored conditions, and the response history in all active degrees-of-freedom is plotted. Statistical analysis is also carried out on the dynamic tension variation in various mooring lines. Through this analysis, a mooring line under maximum tension variation is selected to carry out a failure analysis. Those mooring lines, which showed a maximum mean value in the tension variations, are hypothetically disconnected from the platform to examine its response under such postulated failure conditions. Statistical analysis is carried out on the motion response of the semi-submersible under both catenary and taut-moored configurations for different wave heading angles. It is done for both conditions; namely, moorings are intact in position and under the postulated failure condition. Based on the highest mean value of the response in a particular degree-of-freedom under a specific wave heading angle, the response history of the platform is also plotted. Tables 2.8 and 2.9 summarize the statistical values of the

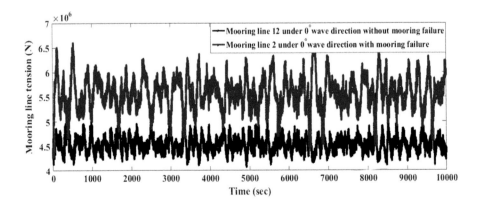

FIGURE 2.15   Tension in taut-mooring lines #12 and #2.

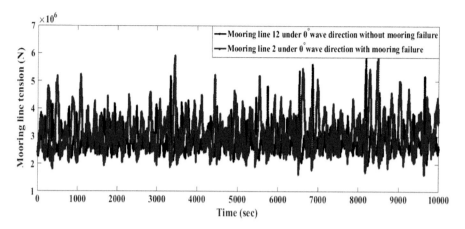

FIGURE 2.16   Tension in catenary mooring lines #12 and #2.

response, under intact and postulated failure conditions of both the mooring configurations; the highest mean values in each response motion are highlighted. Figures 2.8 and 2.9 show the response histories of the semi-submersible under both the mooring layouts for a 10-year return period. Responses are plotted in surge and pitch motion, under both intact and postulated failure conditions, respectively. On each plot, both the responses are over-laid to enable a legible comparison.

It is observed that the mean surge response without failure of the catenary mooring is about 48 m, which is 2.5% of that of the water depth. It is further reduced to about 38.4 m, in the case of a taut-moored system. Both the surge responses are well within the permissible offset values of the compliant offshore platforms deployed for drilling (DNV OS E301). However, by inducing the postulated failure of the selected mooring lines, the mean of the surge response is drifted toward the wave heading side (as, seen in the table, it is −16.5 m). It means that the postulated failure is inducing snap loads to the platform, which can challenge its operational comfort. In case of a taut-moored system, this case does not arise due to a large initial pre-tension imposed on the mooring. It can be seen that the response is still toward a positive drift of about 22 m, as the loads are shared by the adjacent moorings.

By comparing both the tables shown above, it is seen that the heave response of the platform under a catenary moored condition is more than that of taut-moored. As semi-submersibles are subjected to large payloads, both static and dynamic, heave responses are higher than other conventional platforms. In case of a gentle comparison of the response of a semi-submersible with that of a taut-moored tension leg platform (TLP), heave responses in the latter are relatively much less. It is due to the fact that semi-submersibles alleviate the payloads only by positive-buoyant capacity and not by taut-mooring systems. Therefore, it becomes more important to assess its behavior under failure of mooring lines. As explained before, semi-submersibles behave as surface vessels under the failure of mooring lines, making them unsuitable for production drilling. But exploratory drilling up to a certain depth can be carried out. It is also important to note that semi-submersibles

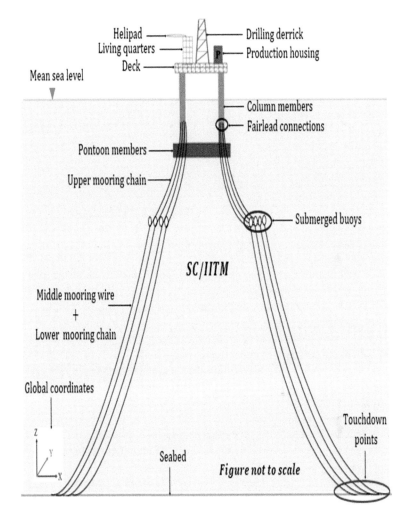

FIGURE 2.17 Semi-submersible with submerged buoy attached to mooring.

are as stiff as other complaint platforms like TLPs in the vertical plane. It is evident from the natural heave period of the platform (see Table 2.7), which is much higher than that of TLPs; normally, heave periods of TLPs are in the range of 2 to 5 s.

The pitch response of the platform, because of a large displacement, is very marginal, even under the postulated failure conditions. It re-emphasizes the fact that semi-submersibles are designed to behave stiffly in rotational degrees-of-freedom. It is achieved in two ways: (1) due to excessive buoyancy, and (2) due to large deck size. Similarly to other compliant offshore structures, semi-submersibles possess a strong coupling between various motion responses. To understand this, let us try to diagnose the power-spectral density (PSD) plots of various motions of the platform. Figures 2.10, 2.11, and 2.12 show the PSD plots of surge, heave, and pitch motion, respectively, under different wave headings for a 10 year return period.

TABLE 2.12

Description of a submerged buoy

Description	Value	Units
Diameter	5.0	m
Structural mass	5861.6	kg
Displaced mass of water	67086	kg
Added mass	33543	kg
Coefficient of drag (C_d)	0.3	NA
Coefficient of inertia (C_m)	0.5	NA

Power spectral density plots of surge response showed major peaks, closer to that of its own natural frequency. This is true in both the mooring system configurations, which means that the type of mooring configuration does not influence the surge response, significantly. But taut-moored systems showed a lesser response in comparison to the catenary-moored platform. It is also interesting to note that peak frequencies of surge response, even under the postulated failure condition, are well away from that of the dominant wave frequency, confirming no resonance effect. It is achieved by the geometric design of the semi-submersible. Hence, compliant and floating offshore structures are termed as a form-dominant design. It is the geometric form that helps to counteract the encountered lateral loads, and not the strength of the members or materials. It is quite unusual in design practices, and therefore is a novelty of design of offshore compliant structures, in general, and semi-submersibles, in particular.

Power spectral density plots of the heave response showed major peaks in closer proximity to its own natural frequency, and that of the pitch frequency as well. This behavior indicates that the platform exhibits a strong coupling between the heave and pitch motions. In physical terms, it means that the heave motion of a significant magnitude will also result in a noticeable pitch response and vice-versa. Further, secondary peaks in the heave response are also observed at a closer vicinity to that of peak wave frequency. It is an alarming behavior as the heave response is influenced by the wave climate. One of the important reasons for such behavior is due to the variable submergence effect, caused by large wave amplitudes. Peak responses in the power spectral density plots of the pitch response are observed in

TABLE 2.13

Configuration of a spread-mooring system

Water depth	Length of the mooring system (m)		
	Upper chain	Middle wire	Lower chain
1500 m	300	2000	1500
2000 m	500	3000	1500

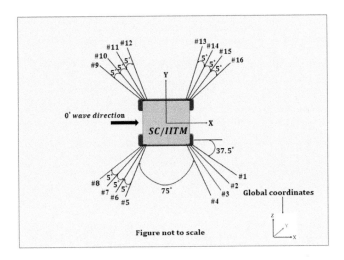

FIGURE 2.18   16-point spread-mooring configuration.

proximity to the natural frequency of pitch motion. Secondary peaks are observed closer to that of heave and surge, indicating a strong coupling of pitch with these motion responses. The presence of secondary responses nearer to that of the wave frequency shows its sensitivity to the wave climate. It is due to such reasons that all floating platforms, in general, and semi-submersibles, in particular, are advised to remain operational only under certain wave conditions; under exceedance of these, they are parked or remain non-operational.

2.2.5 DYNAMIC TENSION VARIATION

The motion response of the semi-submersible, as described in the above section, indicates a strong coupling between various degrees-of-freedom. In addition, it also confirms the influence of the wave climate, heading angle, and postulated failure of mooring lines upon the response of the platform. It is imperative to state that such large motion in flexible degrees-of-freedom will result in dynamic tension variations in the mooring lines. It may be true in both the cases, namely, catenary and taut-moored configurations. As such, semi-submersibles are positive buoyant, and the postulated failure cannot (in fact, should not) result in the structural instability of the platform. In case of eventual failure of any one or more mooring lines, the platform will behave similar to that of a surface vessel and remain afloat. But, as it cannot perform production drilling anymore under such conditions, it may be attributed to a functional failure.

Under such conditions, examining the dynamic tension variation in the mooring lines will lead to their fatigue assessment, and subsequently to a service life estimate of the platform. It should be noted that the failure of mooring lines are tagged to the service life of the semi-submersible as they behave as surface vessels under such failure conditions. In this section, we will examine the insight of the tension variations in the mooring lines, for both the catenary and taut-moored layouts. The

TABLE 2.14

Structural properties of a spread-mooring system

Description	Upper chain	Middle wire	Bottom chain
Mass per unit length (kg/m)	163.86	36.41	163.86
Axial stiffness (MN/m)	676.81	833.91	676.81
Equivalent diameter (m)	0.095	0.095	0.095
Longitudinal drag coefficient	1.15	0.025	1.15
Transverse drag coefficient	2.4	1.6	2.4
Added mass coefficient	1.0	1.0	1.0

fatigue life assessment will be dealt with in the subsequent chapter, with more clarity. For the mooring layout shown in Figure 2.6, all mooring lines may not get excited by the same magnitude. There may be doubt in one's mind why there will be a variation in the tension variation for a symmetric mooring layout. Kindly note that the top side load acting on a semi-submersible is not mass-concentric; it causes eccentricity in terms of the topside load. Further, the geometric layout of a top side consists of the living quarters, drilling derrick, and other accessories, mechanical and electrical equipment, and a helipad. All the items listed above are not placed symmetrically with respect to the center of gravity of the semi-submersible (please see Figures 2.1, 2.2, and 2.3, for details). Therefore, even for a symmetric mooring layout, under the given head-sea conditions, tension in all the mooring lines need not be same; tension variation also changes dynamically. Table 2.10 summarizes the statistics of tension variations in all the twelve mooring lines, under both the layouts for different wave-heading angles.

It is seen from the table that the mean mooring tension for mooring lines #1 and #10 is almost zero, indicating the fact that these mooring lines have undergone postulated failure. Hence, adjacent mooring lines #2 and #11 will be subjected to an additional load, which is now re-distributed from the failed mooring lines #1

TABLE 2.15

Natural periods and damping ratios (2000 m water depth)

Motion	Mooring without buoy		Mooring with buoy	
	T_n (sec)	ξ (%)	T_n (s)	ξ (%)
Surge	208	6.15	195	6.56
Sway	165	6.97	160	6.77
Heave	21	2.47	21	1.20
Roll	24	2.97	23	1.44
Pith	25	0.93	24	1.32
Yaw	49	6.50	54	10.36

TABLE 2.16

Motion statistics of a semi-submersible under intact mooring (10 Y, 2000 m)

Motion	Statistics	Mooring lines without buoy			Mooring lines with buoy		
		0°	45°	90°	0°	45°	90°
Surge (m)	Min	9.65	6.38	−0.006	13.83	8.27	−0.008
	Max	64.17	30.45	0.01	73.53	34.89	0.01
	Mean	34.17	17.12	0.0006	40.2	20.2	0.0006
	Std Dev	8.25	3.27	0.002	9.12	3.6	0.002
Sway (m)	Min	-0.6	1.6	0.97	−0.67	2.51	1.95
	Max	0.03	18.44	24.6	0.03	20.5	27.74
	Mean	−0.14	9.29	11.82	−0.17	10.85	13.81
	Std Dev	0.08	2.15	3.35	0.1	2.36	3.64
Heave (m)	Min	−15.26	−13.81	−13	−14.68	−13.26	−12.25
	Max	−0.78	−2.38	−3.94	0.55	−0.84	−2.74
	Mean	−9.21	−9.21	−9.24	8.22	−8.22	−8.24
	Std Dev	1.17	1.18	1.05	1.27	1.3	1.09
Roll (°)	Min	−0.04	−3.35	−5.81	−0.03	−3.14	−5.67
	Max	0.03	4	6.05	0.04	3.91	6
	Mean	0.003	0.19	0.2	0.003	0.2	0.21
	Std Dev	0.005	0.86	1.4	0.006	0.85	1.39
Pitch (°)	Min	−10.8	−6.41	−0.004	−9.96	−5.85	−0.005
	Max	6.15	4.12	0.004	5.72	4.3	0.004
	Mean	−0.31	−0.11	−5.3E−05	−0.32	−0.12	−4.9E−05
	Std Dev	1.55	1.07	9.4E−05	1.63	1.12	0.001
Yaw (°)	Min	−0.03	−3.62	−0.01	−0.04	−4.63	−0.01
	Max	0.06	6.22	0.01	0.07	6.47	0.01
	Mean	0.001	0.3	−0.001	0.002	0.41	−0.001
	Std Dev	0.009	1.05	0.002	0.01	1.16	0.003
Surge (m)	Min	18.54	9.21	−0.001	22.09	10.8	−0.0009
	Max	125.56	88.53	2.7	128.49	88.54	44.42
	Mean	92.61	74.25	2.65	93	72.33	41.57
	Std Dev	9.86	5.27	0.14	10.76	5.45	2.37
Sway (m)	Min	−0.1	4.49	−34.4	−0.13	5.23	−0.11
	Max	1.31	64.62	8.42	1.14	63.18	26.66
	Mean	0.86	55.53	−22.59	0.64	52.84	12.07
	Std Dev	0.17	3.72	3.81	0.19	3.81	3.91
Heave (m)	Min	−15.53	−15.22	−12.46	−14.94	−14.78	−11.97
	Max	−1.62	−3.12	3.3	−0.65	−1.79	−2.19
	Mean	−8.66	−8.65	−8.67	−7.8	−7.78	−7.81
	Std Dev	1.21	1.24	1.06	1.33	1.39	1.12
Roll (°)	Min	−0.01	−5.01	−5.08	−0.03	−4.5	−5.74
	Max	0.08	2.77	7.35	0.08	2.9	5.95
	Mean	0.02	−0.96	1.36	0.02	−0.72	0.16

(Continued)

TABLE 2.16 Motion statistics of a semi-submersible under intact mooring (10 Y, 2000 m) (*Continued*)

Motion	Statistics	Mooring lines without buoy			Mooring lines with buoy		
		0°	45°	90°	0°	45°	90°
	Std Dev	0.01	0.86	1.42	0.01	0.85	1.39
Pitch (°)	Min	−7.15	−4.12	−0.03	−7.4	−4.32	−0.11
	Max	7.59	6.02	0.1	7.57	5.68	2.15
	Mean	0.7	0.86	0.02	0.47	0.64	0.74
	Std Dev	1.58	1.12	0.008	1.67	1.16	0.07
Yaw (°)	Min	−0.54	−4.7	−0.41	−0.63	−4.94	−0.0016
	Max	0.0015	6.66	−0.0012	0.0001	7.48	1.32
	Mean	−0.31	0.22	−0.26	−0.32	0.34	0.79
	Std Dev	0.05	1.12	0.01	0.05	1.25	0.08

and #10, respectively. Figures 2.13 and 2.14 show the tension variation time history of lines (#11, #2) under catenary and taut-moored layout, respectively. Both the lines are adjacent to the failed mooring lines (#10, #1), respectively. As both the mooring lines (#11, #2) occupy the same position with respect to the failed lines, Figure 2.13 shows the increase in the tension upon the failure of the adjacent line. It can be seen that there is a significant increase in the tension magnitude and tension variation cycles over a period. But this is not true in the case of a taut-moored configuration. A clear and distinct shift, as seen in Figure 2.13, is found missing in Figure 2.14. It is due to the high initial pre-tension in the lines as the latter are taut-moored.

Figures 2.15 and 2.16 show the tension variation time history of lines (#12, #2) under catenary and taut-moored layouts, respectively. Mooring line #12 is connected to the same coordinate where line #10 is connected but failed. On the other hand, line #2 is adjacent to the failed line #1. It is rather interesting to compare the responses of line #12 without the failure of line #10, but line #2 under the failed condition of line #1. Tension variation in mooring line #2 shows a significant tension variation and shift in the magnitude compared to that of mooring line #12. Now, this can be easily attributed to the load sharing that occurred due to the postulated failure of mooring line #1. It is imperative to note that no mooring lines failed due to the load level's exceedance beyond their material capacity. Only postulated failure is intuited to illustrate the motion response of the semi-submersible. It will be instead investigating the fatigue life of the mooring lines under such postulated conditions, which is presented in the next chapter, in detail. In general, a realistic environmental condition, for example, by employing wave spreading and including the effects of vortex-induced vibration (VIV), may result in better accuracy (Hong and Shah, 2018; Hover et al., 2001; Khalak and Williamson, 1991; King, 1977; Leira et al., 2005; Modarres et al., 2011). Further, corrosion in the mooring lines will also influence fatigue life, which is ignored in the current discussion. Because mooring lines are under dynamic tension variation, causing cyclic loading, there will be a higher energy release through the mooring chains. It may also result in further deterioration due to creep, which needs further investigation; it is humbly admitted that these factors are underscored in the current discussion.

2.3 EXAMPLE CASE STUDY 2: NUMERICAL RESPONSE ANALYSIS OF A SEMI-SUBMERSIBLE WITH SUBMERGED BUOY

Massive displacements of a semi-submersible are counter-effects of the viscous forces in the splash zone, which also affects the mean horizontal drift force (Dev and Pinkster, 1995). One of the methods to reduce a semi-submersible's large displacements is by deploying submerged buoys in the mooring lines (Mavrakos et al., 1996; Mavrakos and Chatjigeorgiou, 1997). Numerical investigations have shown that the maximum reduction in tension in the mooring lines should be at the location of the submerged buoy. Besides, it also contributes to the increase in buoyancy to the semi-submersible. Size, location, and the number of submerged buoys to be included in the mooring line need a careful study (Ormberg and Larsen, 1998). Moorings attached with submerged buoys offer excellent resistance to the platform motion. Forces in the mooring system, attached with a buoy, are lesser than that of the mooring system without a buoy. It also enhances the fatigue life of the mooring system (Yan et al., 2018).

The motion response of semi-submersibles can also be controlled by regulating other parameters, namely, the tension-dip angle, pre-tension, configuration, and the number of mooring lines (Chen et al., 2011). The angle between the mooring lines is found to be a key parameter for the platform's response control. Stansberg (2008) experimentally investigated the effects of current-induced forces on the semi-submersibles and showed an increase in the drift forces, mean offset, and motion response. It is found to be responsible for the rise in the tension in the mooring lines. Yan et al. (2016) proposed a mooring system with a submerged buoy for a semi-submersible operating in deep water. They observed that the damping offered by the spread catenary mooring without buoy and with the submerged buoy is nearly the same for the slow-drift motions. Tension at the attachment point of the buoy in the spread catenary mooring was found to be higher. The initial inclined angle and sag-to-span ratio have a significant effect on the motion response of floating platforms and tension in the mooring lines (Wang et al., 2018).

2.3.1 DESCRIPTION OF THE PLATFORM

In the present case, the coupled dynamic analysis of a deep-water semi-submersible pegged with a new configuration of the spread mooring system is discussed. Mooring lines are attached with submerged buoys, as shown in Figure 2.17. Submerged buoys, one in each mooring line, are connected between the upper and the middle section of the mooring, as shown in the figure. Please note that the buoys are attached at about one-third of the water depth. As discussed in the earlier sections, mooring lines offer only position-restraints to the platform. They do not

TABLE 2.17

Motion statistics of a semi-submersible under postulate failure (10 Y, 2000m)

Motion	Statistics	Mooring lines without buoy			Mooring lines with buoy		
		0°	45°	90°	0°	45°	90°
Surge (m)	Min	18.54	9.21	−0.001	22.09	10.8	−0.0009
	Max	125.56	88.53	2.7	128.49	88.54	44.42
	Mean	92.61	74.25	2.65	93	72.33	41.57
	Std Dev	9.86	5.27	0.14	10.76	5.45	2.37
Sway (m)	Min	−0.1	4.49	−34.4	−0.13	5.23	−0.11
	Max	1.31	64.62	8.42	1.14	63.18	26.66
	Mean	0.86	55.53	−22.59	0.64	52.84	12.07
	Std Dev	0.17	3.72	3.81	0.19	3.81	3.91
Heave (m)	Min	−15.53	−15.22	−12.46	−14.94	−14.78	−11.97
	Max	−1.62	−3.12	3.3	−0.65	−1.79	−2.19
	Mean	−8.66	−8.65	−8.67	−7.8	−7.78	−7.81
	Std Dev	1.21	1.24	1.06	1.33	1.39	1.12
Roll (°)	Min	−0.01	−5.01	−5.08	−0.03	−4.5	−5.74
	Max	0.08	2.77	7.35	0.08	2.9	5.95
	Mean	0.02	−0.96	1.36	0.02	−0.72	0.16
	Std Dev	0.01	0.86	1.42	0.01	0.85	1.39
Pitch (°)	Min	−7.15	−4.12	0.03	7.4	−4.32	−0.11
	Max	7.59	6.02	0.1	7.57	5.68	2.15
	Mean	0.7	0.86	0.02	0.47	0.64	0.74
	Std Dev	1.58	1.12	0.008	1.67	1.16	0.07
Yaw (°)	Min	−0.54	−4.7	−0.41	−0.63	−4.94	−0.0016
	Max	0.0015	6.66	−0.0012	0.0001	7.48	1.32

contribute to counteracting the lateral loads. Semi-submersibles counter the lateral loads only by their geometric form and positive buoyancy. Hence, even in case of the removal of the mooring lines, semi-submersibles do not become unstable. But, as the mooring lines inhibit axial tension of high magnitude, a sudden release of this force would result in a snap load on the adjacent mooring lines. It may cause a temporary instability to the platform. When attached to the mooring lines, large, submerged buoys can absorb these shocks to a more considerable extent. Further, as these submerged buoys possess immense buoyancy, they will offer position-restraint to the platform even under the failure of the bottom segment of the mooring lines. It may be curious to note why the upper section of a given mooring line will not fail. It is due to the shorter length and closer connectivity to the platform. Bottom segments, connected from the submerged buoys, are very long and highly flexible. They follow a slackened profile, due to which they attract wave loads. Further, because of their slackening effect, they are also susceptible to the attack of marine growth, which corrodes them further. Because of the above reasons, the

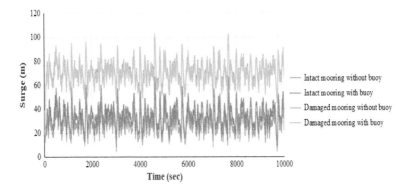

FIGURE 2.19 Surge response of semi-submersible (0°, 10 Y, 2000 m).

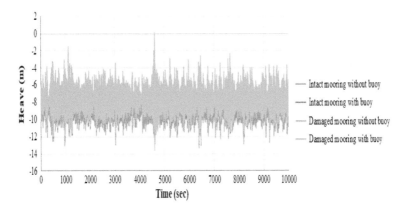

FIGURE 2.20 Heave response of semi-submersible (45°, 10 Y, 2000 m).

bottom segment of the mooring lines, extended beyond the submerged buoys, is more likely to fail than the upper section.

Table 2.11 shows the structural details of the semi-submersible with submerged buoys, which are being analyzed for two different water depths, namely, 1500 and 2000 m.

2.3.2 16-Point Mooring System with Submerged Buoy

Table 2.12 shows the details of submerged buoys connected to the mooring lines; each mooring has one buoy attached to it, as shown in Figure 2.17. Figure 2.18 shows the details of the spread-mooring configuration, used to position-restrain the semi-submersible.

Table 2.13 shows the 16-point spread-mooring configuration, while Table 2.14 shows the structural properties.

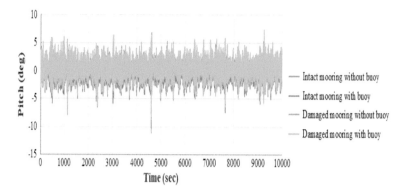

FIGURE 2.21   Pitch response of semi-submersible (0°, 10 Y, 1500 m).

2.3.3 NATURAL PERIODS AND DAMPING

Free-decay tests are carried out on the moored semi-submersible at two different water depths, namely, 1500 and 2000 m. Table 2.15 summarizes the natural periods and damping ratios of the platform with the submerged buoy.

It can be observed that deploying buoys in the mooring lines decrease the natural periods in the compliant degrees-of-freedom. The addition of a submerged buoy makes the platform stiffer, while it also contributes to an additional mass of the platform.

2.3.4 MOTION RESPONSE

A detailed dynamic analysis is carried out under moored conditions, and the response history in all active degrees-of-freedom is plotted. Statistical analysis is also carried out on the dynamic tension variation in various mooring lines attached to the submerged buoys. Those mooring lines, which showed a maximum mean value in

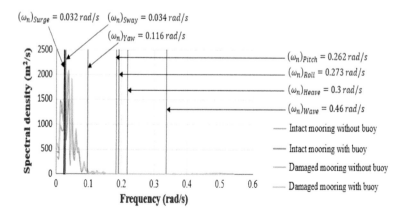

FIGURE 2.22   PSD plots of surge response of semi-submersible (0°, 10 Y, 2000 m).

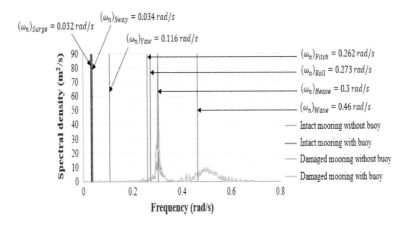

FIGURE 2.23 PSD plots of heave response of semi-submersible (45°, 10 Y, 2000 m).

the tension variations, are hypothetically disconnected from the platform to examine its response under such postulated failure conditions. The response history of the platform is also plotted based on the highest mean value of the response in a particular degree-of-freedom under a specific wave heading angle.

Tables 2.16 and 2.17 summarize the response's statistical values under intact and postulated failure conditions for 2000 m water depth; the highest mean values in each response motion are highlighted. Figures 2.19, 2.20, and 2.21 show the response histories of the semi-submersible with and without attaching the submerged buoys to each mooring line. Based on the maximum average value for each wave heading, responses are plotted for surge, heave, and pitch motion. For the convenience of comparing the responses, mooring with intact and postulated failure conditions are over-laid.

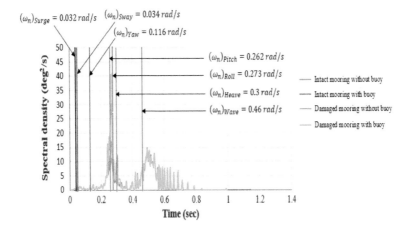

FIGURE 2.24 PSD plots of pitch response of semi-submersible (0°, 10 Y, 2000 m).

TABLE 2.18

Mooring lines under postulated failure (1500 m water depth)

Return period	Mooring without buoy			Mooring with buoy		
	0°	**45°**	**90°**	**0°**	**45°**	**90°**
10 years	# 1	# 4	# 4	# 10	# 9	# 2
100 years	# 14	# 16	# 5	# 12	# 10	# 14
	# 12	# 11	# 2	# 9	# 8	# 5
	# 14	# 12	# 13	# 11	# 9	# 6

TABLE 2.19

Mooring lines under postulated failure (2000 m water depth)

Return period	Mooring without buoy			Mooring with buoy		
	0°	**45°**	**90°**	**0°**	**45°**	**90°**
10 years	# 1	# 13	# 1	# 1	# 13	# 15
100 years	# 14	# 14	# 6	# 14	# 14	# 16
	# 1	# 13	# 2	# 12	# 11	# 8
	# 14	# 14	# 5	# 14	# 12	# 9

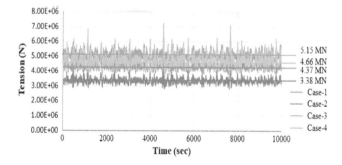

FIGURE 2.25   Tension in the mooring lines (0°, 10 year return period). Case-1: Intact mooring #1 without buoy; Case-2: Intact mooring #15 with buoy; Case-3: Intact mooring #15 without buoy after damage of mooring lines #1 and #14; Case-4: Intact mooring #13 with buoy after damage of mooring lines #10 and #12.

Figures 2.22, 2.23, and 2.24 show the power spectral density plots of the surge, heave, and pitch responses, respectively. The PSD of surge plots shows a significant peak, closer to that of the surge and sway frequencies. No peaks are observed at other frequencies, including that of the wave peak frequency. Surge PSD shows more energy content with a higher magnitude at a lower frequency, cautioning that the platform is more vulnerable to wind loads. The presence of submerged buoys improves the platform's stiffness but does

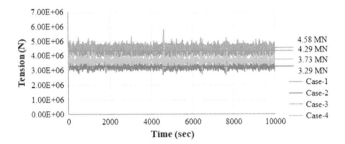

FIGURE 2.26 Tension in the mooring lines (45°, 10-year return period). Case-1: Intact mooring #15 without buoy; Case-2: Intact mooring #13 with buoy; Case-3: Intact mooring #3 without buoy after damage of mooring lines #4 and #16; Case-4: Intact mooring #11 with buoy after damage of mooring lines #9 and #10.

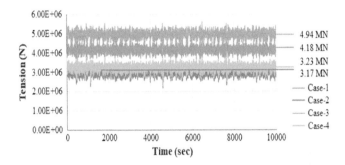

FIGURE 2.27 Tension in the mooring lines (90°, 10-year return period). Case-1: Intact mooring #12 without buoy; Case-2: Intact mooring #9 with buoy; Case-3: Intact mooring #7 without buoy after damage of mooring lines #4 and #5; Case-4: Intact mooring #1 with buoy after damage of mooring lines #2 and #14.

not contribute to disbursement of the lateral loads. It only saves the mooring lines from sudden failure and helps to avoid snap load on the mooring lines.

Heave plots showed major peaks at its frequency and in closer proximity to that of the pitch natural frequency. This indicates a strong coupling between the heave and pitch motions of the semi-submersible. Many secondary peaks of considerable magnitude are observed at the frequency closer to that of the waves. As expected, the platform is sensitive to the wave amplitude and period because it derives the strength to resist lateral loads only from the deep-draft (Hussain et al., 2009). Therefore, higher magnitude waves will result in the variable submergence effect and make the heave motion wave-sensitive. The energy content of the heave plots is lesser than that of the surge, indicating that the platform is very stiff in the vertical plane of motion. Peaks seen at the higher frequencies also confirm this behavior.

The power spectral density plots of the pitch response show major peaks at frequencies of pitch and heave, confirming the strong coupling between these motions. Secondary peaks, seen closer to the peak wave frequency, indicate that the pitch motion is wave-sensitive. But, because of the larger platform geometry and being a very stiff degree-of-freedom, pitch motion does not pose a severe threat to the platform stability. Larger spread PSD plots of the pitch response confirm that the pitch motion is also wind-sensitive.

2.3.5 Tension Variation in Mooring Systems

Significant displacement of the platform in the horizontal plane confirmed through a large magnitude of the surge and sway motion results in the dynamic tension variation in the mooring lines. These tension variations are also sensitive to the wave-heading angle as the symmetry of the mooring configuration is not the same under all approach angles. Tables 2.18 and 2.19 show the summary of the mooring lines under postulated failure. They are chosen based on the maximum tension variations.

Figures 2.25, 2.26, and 2.27 show the tension variation time history, simulated for 10000 secs, which is about a 3-hour storm duration. Plots are shown for different wave-heading angles of 0°, 45°, and 90°, respectively. The mean values of the tension, in several cases, are indicated for more clarity. Failed mooring lines can be correlated with the above tables for a better understanding of the failure cases.

It is seen from the above figures that tension from the failed mooring lines is transferred to the adjacent moorings. It is also observed that the presence of a submerged buoy reduced the amplitude of the tension, but does not influence the number of stress cycles. However, reducing the magnitude of stress intensity enhances the fatigue life of the mooring lines and, thereby, the platform. Under the varying tension in the mooring lines, fatigue life needs to be estimated as they undergo a large number of stress cycles. Detailed fatigue analysis is discussed in the next chapter.

3 Reliability and Fatigue Life

3.1 INTRODUCTION

Oil exploration has a history that goes back to the 19th century; the first offshore platform was commissioned in 1924. It was a fixed platform, offering resistance in all six degrees-of-freedom. Subsequently, the platform's rigidity in a few degrees-of-freedom is released to reduce the forces that arise from wave-structure interaction; hence, offshore platforms became compliant. However, upon total exploitation of oil reserves in shallow and intermediate water depths, the installation of offshore platforms moved toward deep and ultra-deep waters. A significant component of the sub-structure was replaced with tethers/mooring lines, leading to a dual advantage: cost reduction and higher flexibility (Reddy and Swamidas, 2013; Chandrasekaran and Nagavinothini, 2020; Chandrasekaran, 2019a,b; Chandrasekaran and Manda, 2019; 2020; Chandrasekaran et al., 2015; 2013a–c; 2011; Chandrasekaran and Mayank, 2017; Chandrasekaran and Yuvraj, 2013).

Structural forms of the complaint offshore platforms are conceived by enabling higher flexibility in the horizontal plane, but with rigidity in the vertical plane (Chandrasekaran and Madhuri, 2012a,b; Chandrasekaran and Thailammai, 2018). The later was achievable by taut-moored tethers with high axial pretension. Recent developments of offshore platforms for deep waters has led to the commissioning of the Single Point Anchor Reservoir (SPAR). It consists of a deep-draft, heavily ballasted central cylinder that supports the deck; the platform is anchored to the seabed by the spread-mooring system (Chandrasekaran et al., 2010; 2015; 2008; Chandrasekaran and Gaurav, 2008; Chandrasekaran et al., 2007a–d). Though its geometric form enables hydro-dynamic stability, large displacements of the deck and limited payload capacity are serious concerns that have remained unaddressed for a long time (Chandrasekaran and Nannaware, 2013; Chandrasekaran and Nassery, 2015; 2017; Chandrasekaran et al., 2006a,b; Chandrasekaran et al., 2004; Chandrasekaran and Jain, 2002a,b). New-generation platforms, namely, the offshore Triceratops, have a novel design, as they possess both better station-keeping and motion characteristics (Chandrasekaran and Lognath, 2017; 2015; Chandrasekaran and Madhuri, 2015; 2013a; 2012; 2012a,b). Recent research on the response behavior of Triceratops confirmed that the deck is capable of maintaining the tranquility even under rough seas (Chandrasekaran and Nagavinothini, 2017; 2018a,b; 2019a–c; 2020a–d; 2019e; Chandrasekaran and Jamshed, 2015; 2017).

One of the most successful platforms in deep waters, the tension leg platform (TLP), possesses a more significant draft that makes the mooring lines always in high axial pretension (Chandrasekaran and Kiran, 2017; 2018; Prucz and Soong, 1984; Chandrasekaran and Jain, 2002a,b). The taut-mooring system of TLPs makes the design adaptable to deep waters, but also makes it essential for predicting the probability of failure under fatigue behavior (Chandrasekaran et al., 2006a, b; Chandrasekaran and Yuvraj, 2013). Catastrophic failure occurs when the response exceeds a threshold during extreme storm waves (Chandrasekaran and Thailammai, 2016; 2018; Chandrasekaran, 2019b; El-gamal et al., 2014). While the lower limit is considered the initial pretension level, higher impulsive forces act on the legs, which makes its behavior uncertain. The mooring system is deemed to have failed when the legs at the corner fail, which subsequently initiates a postulated failure to the platform.

The continuous use of the concepts of reliability-based design of offshore structures includes the benefits of probabilistic modeling, which maps the modeling uncertainties (Bjerager, 1990; Chandrasekaran, 2016a,b; 2014). Reliability and the probability of failure of a structure are the converse of each other. While the probability of failure involves many members failing and includes the modes of failure, reliability is mapped from the failure of the critical members only. In the Monte Carlo simulation, when the probability of failure lies in the range (10^{-3} to 10^{-10}) or even a lesser order, only a few points will fall under the failure domain resulting in a zero probability of failure (Melchers, 1988; Chandrasekaran, 2015; 2016a,b). If a large number of simulations are considered, then the computational effort on the limit state function increases along with the computational time. Therefore, it is essential to generate the random numbers through a sampling density function, which has the maximum likelihood enclosed within the limit state function itself. An indicator function will, therefore, provide information about whether the generated point lies within the failure domain or the safe domain. In general, sampling from the failure domain is considered a biased sampling, but dividing it with the original density function will make it unbiased (Ranganathan, 1999; Onoufriou, 1999).

The First-Order Reliability Method (FORM), Second-Order Reliability Method (SORM), and Monte Carlo Simulations (MCS), which are employed in the reliability analyses, are often complementary to each other. Problems handled with MCS can also be handled with FORM/SORM, and vice-versa. For example, in the engineering problems, the failure probability will vary in the range (10^{-4} to 10^{-8}), for which MCS is not beneficiary. It is because it is computationally inefficient for such orders of failure probability. On the other hand, problems involving discrete distributions and the transformation of one space variable to another do not exist in FORM/SORM but can be solved by Monte Carlo methods. However, the convergence in the Monte Carlo method depends on the order of the probability of failure. The efficiency of the FORM/SORM is independent of the probability of failure but decreases with the increase in the number of random variables. However, the effectiveness of MCS is improved with the sampling technique, which can reduce the variance to zero (theoretically) and further improved by adaptive sampling, where sampling density is per some decision-making process. It is interesting to note that the convergence of both methods depends on the initial design points.

Monte Carlo simulation is one of the statistical methods developed to relate the probabilistic values and the respective vital parameters. It is one of the significant components of the risk analysis, carried out in parallel with (or even substitutes for) the decision tree analysis. Risk is a qualitative phenomenon, while probability is quantitative. Various uncertainties that arise from the environment loads and material characteristics under aging can be quantified by employing a proper distribution to an acceptable level of accuracy (Karmazinova and Melcher, 2012; Sadowski et al., 2015). While the risk assessment is a mathematical process of quantifying the potential loss of assets or changes occurring in the due course of time, computing the probability of an event is one of the forms of risk analysis (Skrzypczak et al., 2017).

Environmental loads on offshore structures follow a random process, which is usually characterized by a probability distribution. It is further characterized by a density function with two or more parameters, namely, the mean and standard deviation. Monte Carlo simulation generates samples from these distributions and estimates the variable, which is dependent on the random variable. The main objective of Monte Carlo simulation is to obtain the probability distribution of the dependent variable, which otherwise is complicated to solve analytically. Alternatively, a sensitivity analysis is also useful to identify the most influencing random variables in the Monte Carlo simulation process. Despite all these advantages, there are a few drawbacks: (i) the output depends on the modeling method; (ii) the processing number of iterations converge, and (iii) it takes a considerable amount of time, which makes it computationally inefficient.

3.2 EXAMPLE STUDY: TRICERATOPS

Possessing the advantages of both the platforms, namely, TLP and SPAR, the offshore Triceratops is observed as an adaptable platform for ultra-deep water due to its novel geometric form. The offshore Triceratops has a reduced carbon footprint with an improved economy because of its reduced usage of resources. Under the current stage of design conceptualization, it is imperative to examine Triceratops under postulated failure by removing one or more tethers of the buoyant legs. Fatigue failure, initiated by such conditions, needs to be visited in detail to ascertain the reliability of Triceratops. Codes for design construction practice evolved in time for better and optimistic design by considering the risk and reliability. Concepts of the reliability account for the random variables with a probability distribution of vital parameters, namely, the mean and standard deviation. These concepts are useful to calibrate the developed code of practice.

The dynamic response of Triceratops under various environmental loads showed that even the heave response, which is a stiff degree-of-freedom, is influenced by the change in the pretension of tethers (Chandrasekaran and Jamshed, 2017a, b; 2015; Chandrasekaran et al., 2015a–c). Ball joints connecting the deck and buoyant legs help maintain the deck (Chandrasekaran and Nagavinothini, 2018a–c). Triceratops is compliant in the horizontal plane and stiff in the vertical plane due to which it showed a lesser response of the deck compared to that of the buoyant legs. This is attributed to the partial isolation extended by the ball joints. They restrain the transfer of rotation from the legs to the deck or vice-versa. However, under the eccentric loading on the deck, the Triceratops is observed to be under Mathieu instability, leading to the fatigue failure of tethers (Chandrasekaran and Kiran, 2018a,b; 2017; Det Norske Veritas, 2016a,b; 2008).

The deck of the platform is supported by three buoyant legs, which are, in turn, connected to taut-moored tethers for position-restraining the platform. Each buoyant leg is restrained in position by a set of three taut-moored tethers. Ball joints that connect the buoyant legs to the deck partially isolate the deck. They restrain the transfer of rotation from the legs to the deck and thus provide tranquility for drilling and exploration operations. As buoyancy exceeds the weight of the platform, tethers will always remain in axial tension. No slackening occurs, but relaxation in the pre-tension is expected under large displacements of the platform. It, therefore, results in a reversal of axial forces in tethers, initiating a fatigue failure. Because of this phe-nomenon, axial tensile stresses in tethers will vary about a mean value, which will lead to crack initiation. It will further lead to crack propagation, and ultimately, to fatigue.

Moorings/tethers are fabricated in segments, and they are considered to be failed when one (or all) of the joints fail (Prucz and Soong, 1984; Sadowski et al., 2015). Fatigue damage to tethers is generally computed by performing the rainflow counting and subsequently applying Miner's rule to obtain the linear damage (Marsh et al., 2016; Nagavinothini and Chandrasekaran, 2019a,b; 2020). A bi-linear S-N curve with slopes of 3 and 5 is assumed from the code (Det Norske Veritas, 2016a,b); rainflow counting is employed to compute the range of stress cycles and the corresponding cycle averages. The Goodman diagram is a useful tool that can be used to convert the non-zero mean stress process to a zero-mean process (Marsh et al., 2016).

3.2.1 DESCRIPTION OF THE PLATFORM

A larger deck and the complex joints of compliant structures like SPARs (Single Point Anchor Reservoirs) and TLPs (Tension Leg Platforms) will make them expensive for exploration in ultra-deep waters. Further, it is challenging to inspect and maintain these systems (Chandrasekaran and Gaurav, 2017; 2008; Chandrasekaran, 2019a,b; 2018; 2017; 2015a,b; Chandrasekaran and Ajesh, 2019; 2017). Alternatively, Triceratops exhibits better motion characteristics and station-keeping (Nagavinothini and Chandrasekaran, 2019a; Chandrasekaran and Nagavinothini, 2018a,b; Chandrasekaran and Madhuri, 2015; 2013). The structural weight counteracts buoyancy generated due to the deep draft of buoyant legs, and the excess is shared with the tethers as pretension. Detailed studies carried out on the response behavior of Triceratops showed that the deck response is relatively lower than that of the buoyant legs. Because of the partial isolation they impose, the ball joints reduce both the translation and rotation of the deck even under rough sea states (Chandrasekaran, 2016a; Chandrasekaran and Nagavinothini, 2017a,b; Chandrasekaran and Jamshed, 2017a,b; 2015). An increase in Pre-tension of tethers restricts the movement of buoyant legs, reducing the natural periods of the surge and sway by 70% (Chandrasekaran et al., 2015a–c). Extreme storm waves can induce considerable fatigue damage in critical joints. In the presence of such waves, a maximum surge and sway response is observed in different wave-headings (Chandrasekaran and Yuvraj, 2013). The presence of current offsets the surge response and influences the heave response (Chandrasekaran and Nagavinothini, 2018a).

The platform's preliminary design confirmed its suitability in ultra-deep waters successfully (Chandrasekaran and Nagavinothini, 2018b; Chandrasekaran and Mayank, 2017). However, a higher order of compliancy in the horizontal plane causes more tension variations in the tethers under increased payloads, which can result in Mathieu instability (Chandrasekaran and Jain, 2016; Chandrasekaran et al., 2009; Chandrasekaran and Kiran, 2018; Chandak and Chandrasekaran, 2009; Chandrasekaran et al., 2008a,b). The presence of eccentric loads invokes Mathieu instability and reduces the fatigue life of the Tethers (Chandrasekaran and Kiran, 2018a,b; 2017). Because of the asymmetric form of waves, the structure is prone to a ringing response, which will induce fatigue damage in tethers (Chandrasekaran and Jamshed, 2017a,b; 2015). Tethers undergo different stress fluctuations depending on the position in a given sea state. Therefore, a detailed fatigue life estimate of tethers for different wave, wind, and current conditions are demanded, and the reliability studies are necessary. Figure 3.1 shows the numerical model of Triceratops. Figure 3.2 shows the plan view of the model. The deck is modeled as a solid body having six-degrees-of-freedom. The deck comprises triangular and quadrilateral plate elements with appropriate mass properties. Ball joints connect the deck and the buoyant legs, which partially isolates the deck from the legs. Buoyant legs qualify for Morison elements (D/L < 0.2) and are therefore modeled as line bodies. From the keel of the buoyant legs, tethers are connected to the seabed with an initial pretension. The Pierson Moskowitz (PM) spectrum is used to model the random waves for different sea states, both under normal and postulated failure conditions of the tethers. A rainflow counting algorithm is used to obtain the cycle ranges and the corresponding cycle averages. A non-zero mean stress process is converted into a zero-mean process, which is necessary to apply Miner's rule. Fatigue life, thus estimated, is subsequently mapped to the reliability index using nonlinear regression. Table 3.1 shows the structural details of the platform considered for the present example study.

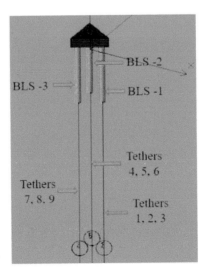

FIGURE 3.1   Numerical model of Triceratops.

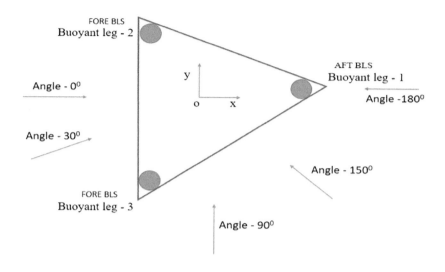

FIGURE 3.2   Plan view of the model.

TABLE 3.1
Structural details of Triceratops

Description	Value
Side of the triangular deck	95.0 m
Water depth (d)	1055 m
Length of tether (L0)	900.76 m
Cross-sectional area of tether	0.213 m^2
Axial stiffness of tether	49.7 MN/m
Number of tethers per each leg	3
Number of buoyant legs	3
Diameter of buoyant leg	15.0 m
Length of buoyant leg	174.24 m
Thickness of buoyant leg	40 mm
Buoyancy per leg	274 MN
Weight of deck + payload	262 MN
Total weight of buoyant leg and tethers	209 MN
Total weight of ballast	91.5 MN
Pre-tension in tether (T0)	28.8 MN
Metacentric height of buoyant leg	37.25 m
Distance between mass Center (CG) and Center of Buoyancy (COB) of buoyant leg	−37.163 m

3.3 ENVIRONMENTAL FORCES

Ocean waves are stochastic; hence, the statistical values of the wave heights will be the same for the wave records recorded in a small interval of time. The Pierson-Moskowitz (PM) spectrum is considered for the generation of random waves. Triceratops is subjected to different sea states under different wave-headings in both the normal and failure conditions of the tethers. Table 3.2 summarizes the details of sea states used in the example study. The PM spectrum is a one-parameter spectrum in which waves are generated due to the wind action. The winds are considered unlimited fetch, and they are assumed to blow for a longer duration. The following relationship gives the PM spectrum model:

$$S(\omega) = \frac{0.0081g^2}{\omega^5} \times \frac{1}{e^{0.74*\left[\frac{\omega U_w}{g}\right]^{-4}}} \tag{3.1}$$

Alternatively, in terms of the spectral peak frequency, the PM spectrum model is expressed as follows:

$$S(\omega) = \frac{0.0081g^2}{\omega^5} \times \frac{1}{e^{1.25\left[\frac{\omega}{\omega_0}\right]^{-4}}} \tag{3.2}$$

where ω is the angular frequency and ω_0 is the angular frequency of the spectral peak. The variance of the wave elevation (σ^2) is equal to the area under the spectral curve, indicated as the zeroth moment (m_0) and is given by the following relationship:

$$\sigma^2 = m_0 = \int_0^\infty S(\omega)d\omega \tag{3.3}$$

$$S(\omega) = 5\sigma^2\left[\frac{\omega^{-5}}{\omega_0^{-4}}\right] \times \frac{1}{e^{1.25\left[\frac{\omega}{\omega_0}\right]^{-4}}} \tag{3.4}$$

TABLE 3.2
Sea states used in the study

Sea state	Nomenclature	(H_s, T_z) (m, s)	Wind velocity (m/s)
Low	S_1	(2.88, 5.45)	10
Moderate	S_2	(6.5, 8.18)	15
High	S_3	(10, 10.3)	35
Very high	S_4	(15, 12.98)	45

The Root Mean Square (RMS) of the sea-surface elevation (σ) is related to the peak frequency as given below (Table 3.2):

$$\sigma = \sqrt{\frac{0.0081}{5}} \times \frac{g}{\omega_0^2} \tag{3.5}$$

$$H_s = 4\sigma \tag{3.6}$$

Mostly, surface current velocities vary from 1%–3% of the sustained wind speed (Reddy and Swamidas, 2016). In this example study, surface current velocities 0.12 and 0.16 m/s (indicated as C_1, C_2) are applied for the low and moderate sea states, respectively. A current velocity of 1.13 m/s (shown as C_3) with a return period of about ten years is applied for high and very high sea states (American Petroleum Institute, 2007).

3.4 MOTION RESPONSE

Natural periods in all degrees-of-freedom are obtained by the free decay test. Table 3.3 summarizes the natural period and the damping ratios, both under intact condition and the postulated failure condition of all tethers. It is seen from the table that Triceratops exhibits a high degree-of-compliancy in the horizontal plane, confirmed by long periods in the surge, sway, and yaw motions. But it remains very stiff in the vertical plane as desired for the drilling operation. It is confirmed by very low periods in heave, roll, and pitch motions. Under the failed condition, the platform exhibits a stiff behavior marginally. It is due to the balance of payload only through the deep-draft of buoyant legs. Under design, this is a preferred characteristic; the FORM-dominating property of the platform achieves this. A stiff behavior under free-floating conditions also ensures the safety of the platform.

The time history of the heave response of the deck is shown in Figures 3.3 and 3.4 for a particular sea state (S_1) under 0° and 30° wave-headings, respectively. It is seen

TABLE 3.3
Dynamic properties of the platform

Degrees-of-freedom	Before the failure of 3 tethers		After the failure of 3 tethers	
	Natural period (Tn) (sec)	Damping ratio (ζ) (%)	Natural period (Tn) (sec)	Damping ratio (ζ) (%)
Surge	133.8	0.59–1.59	129.03	1.10–4.80
Sway	133.8	0.29–1.59	128.02	0.23–2.71
Heave	2.2	0.43–1.4	2.56	0.03–0.81
Roll	3.12	0.02–0.12	3.75	0.025–0.169
Pitch	3.14	0.008–0.91	3.69	0.025–0.0769
Yaw	169.07	0.43–4.45	150.43	1.9–4.28

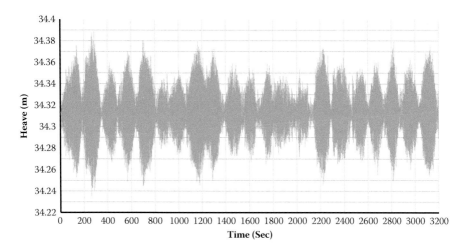

FIGURE 3.3 Heave response of deck (S_1–0°).

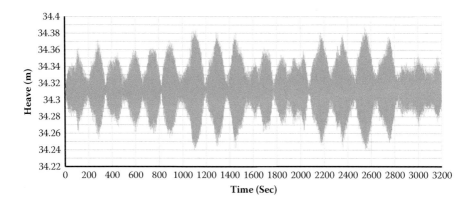

FIGURE 3.4 Heave response of deck (S_2–30°).

from the plots that the heave response of the deck is sensitive to the wave-heading angle. As the waves encounter only the buoyant legs and not the deck directly, the response of the deck is an indirect effect caused by the buoyant legs. As the interaction between the buoyant legs is different under different wave-heading angles for the same sea state, the heave response of the deck is mixed.

The area under the PSD plot, shown in Figure 3.5, is found to be different for different wave-headings. Also, the heave response of the deck is a maximum at a frequency of about 2.84 rad/sec and is not influenced by the wave-heading angles.

3.5 DYNAMIC TENSION VARIATION IN TETHERS

Platform motion under different sea states and wave-heading angles cause dynamic tension variations in the tethers. Results obtained from the statistical analysis of the stress processes are presented in Tables 3.4, 3.5, and 3.6 for different buoyant legs, respectively. The mean, which is equal to the pre-stress value; the median; and the mode are all found to be the same for all the sea states in each tether of the buoyant legs. Skewness and kurtosis values were observed to be almost zero in all the considered sea states, which confirms that the stress-process is symmetric about the mean value. Figure 3.6 shows both the time history and frequency domain plot of

TABLE 3.4

Descriptive statistics of stress process without postulated failure for tethers in buoyant leg-1

Wave condition	Mean	Median	Mode	Standard deviation	Skewness	Kurtosis
S_1-$0°$	133.826	133.814	133.814	9.652281	0.0282	−0.2425
S_1-$30°$	133.826	133.828	133.814	7.04920	0.0116	−0.2803
S_1-$90°$	133.826	133.814	135.663	7.672286	0.0231	−0.2783
S_1-$150°$	133.826	133.785	134.354	7.31928	0.0142	−0.2563
S_1-$180°$	133.827	133.757	133.629	9.865305	0.0098	−0.2430
S_1+C_1-$0°$	133.826	133.814	133.814	10.3588	0.0205	−0.3247
S_2-$0°$	133.825	133.800	133.814	10.07773	0.004	−0.2395
S_2+C_2-$0°$	133.829	133.828	133.828	11.54199	0.0087	−0.0408
S_3-$0°$	133.804	133.757	133.814	10.37294	0.00234	−0.1696
S_3+C_3-$0°$	133.935	134.084	132.249	9.889841	−0.0510	−0.0788
S_4-$0°$	133.731	133.800	131.837	9.188719	−0.0138	0.5261
S_4+C_3-$0°$	133.897	133.928	134.767	16.14269	0.0014	−0.2477

TABLE 3.5

Descriptive statistics of stress process without postulated failure for tethers in buoyant leg-2

Wave condition	Mean	Median	Mode	Standard deviation	Skewness	Kurtosis
S_1-$0°$	133.133	133.103	133.131	6.534801	0.0430	−0.1405
S_1-$30°$	133.132	133.117	133.188	7.359715	0.0084	−0.2633
S_1-$90°$	133.132	133.160	133.188	7.934831	0.0082	−0.2343
S_1-$150°$	133.133	133.103	131.766	6.84544	0.0168	−0.3156
S_1-$180°$	133.134	133.074	129.803	6.628961	0.0235	−0.1360
S_1+C_1-$0°$	133.137	133.131	134.525	6.619546	0.0170	−0.1542
S_2-$0°$	133.139	133.046	132.505	6.082811	0.0588	0.0803
S_2+C_2-$0°$	133.1521	133.131	133.131	8.488369	0.0137	−0.3389
S_3-$0°$	133.129	133.103	132.918	5.746047	0.0435	−0.0721
S_3+C_3-$0°$	133.448	133.401	133.401	6.265833	0.0213	0.1433
S_4-$0°$	133.100	133.046	132.918	5.38623	0.0240	0.4110
S_4+C_3-$0°$	133.456	133.401	133.401	8.623382	0.0094	−0.1952

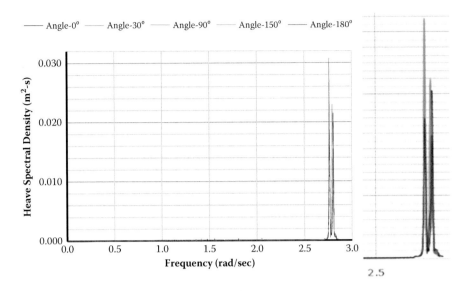

FIGURE 3.5 PSD plots of heave response of deck (S_1).

TABLE 3.6

Descriptive statistics of stress process without postulated failure for tethers in buoyant leg-3

Wave condition	Mean	Median	Mode	Standard deviation	Skewness	Kurtosis
S_1-0°	133.122	133.103	133.814	6.486349	0.0423	−0.1233
S_1-30°	133.123	133.089	133.103	6.756979	0.0218	−0.2656
S_1-90°	133.121	133.117	131.908	7.927679	0.0117	−0.2996
S_1-150°	133.121	133.074	133.757	7.416661	0.0096	−0.2386
S_1-180°	133.123	133.046	130.657	6.58568	0.0237	−0.1511
S_1+C_1-0°	133.126	133.131	132.562	6.598175	0.0156	−0.1600
S_2-0°	133.129	133.046	135.009	6.030103	0.0595	0.0803
S_2+C_2-0°	133.141	133.131	134.354	8.426944	0.0111	−0.3392
S_3-0°	133.118	133.089	133.074	5.831045	0.0433	−0.0257
S_3+C_3-0°	133.438	133.387	131.169	6.206943	0.0254	0.1379
S_4-0°	133.090	133.017	132.505	5.333345	0.0260	0.4045
S_4+C_3-0°	133.446	133.401	136.203	8.610551	0.0093	−0.1602

the tension variation of tethers in one of the buoyant legs under (S_1–0°). It is seen from the figure that the process is following a normal distribution.

3.6 FATIGUE ANALYSIS AND RELIABILITY ASSESSMENT

The generation of ocean waves is an in-deterministic process. When offshore floating platforms counteract such waves, they experience large displacements,

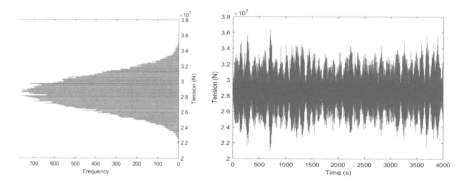

FIGURE 3.6 Dynamic tension variation of tethers under buoyant leg-1 (S_1–0°).

which are restored by tethers or mooring lines. As a result, a variety of tension is induced, which becomes more severe in the presence of current. Because of this phenomenon, the stress amplitudes in the tethers and mooring lines vary about a mean position, which will lead to crack initiation. It further will lead to crack propagation, and subsequently, to fatigue failure. S-N curves are plotted for the fatigue life and are generally used to estimate the fatigue life. A bi-linear S-N curve, adopted in the present study has the following constants:

$$m = 3; \quad log_{10}a = 12.049 \, for \, \sigma \, \leq \, 103 N/mm^2 \qquad (3.7a)$$

$$m = 5; \quad log_{10}a = 16.081 \, for \, \sigma \, > \, 103 \, N/mm^2 \qquad (3.7b)$$

The endurance limit is 41.45 N/mm^2 at 10^8 cycles and is used in the current example study (DNVGL-RP-C203, 2016). To obtain the cycle ranges, and the corresponding cycle averages, rainflow counting is adopted.

As the S-N curve is plotted for a zero-mean process, and the tethers are always in pretension, the cycle average will be equal to the initial pre-stress. Hence, to use the S-N curve for determining the number of cycles that create damage to unity, each cycle should be converted into a zero-mean cycle. The Goodman diagram, shown in Figure 3.7, is used to convert from a non-zero mean process to a zero-mean process.

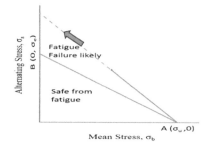

FIGURE 3.7 Goodman diagram.

3.6.1 PALMGREN-MINER RULE

Fatigue is a complex process in which the total strain energy of the 'N' constant-amplitude cycles is equal to the strain energy of 'n' variable-amplitude cycles. The relationship is given as below:

$$FD = \Sigma_{i=0}^{z} \frac{n_i}{N_i} \tag{3.8}$$

where N_i is the number of cycles required to create fatigue damage of unity for a given cycle range, and n_i is the number of cycles observed for the given cycle range.

3.6.2 RELIABILITY AGAINST YIELDING

A member or structure is reliable only when it satisfies the design requirements within its design life. A limit state surface can be linear, plane surface, or even shapeless, and it depends on the random variables involved in assessing the load and resistance. The following expression gives the limit state considered for the failure against yielding.

$$M = R - S \tag{3.9}$$

The reliability index is expressed as the distance between origin and design point in the reduced coordinate system. Figure 3.8 shows the plot of the reliability index of

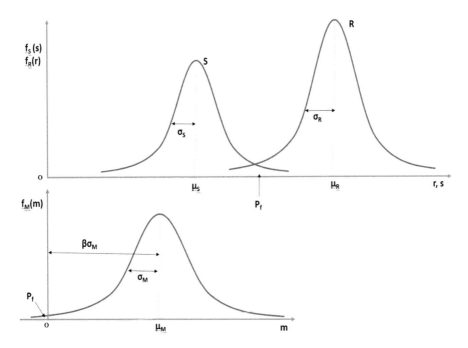

FIGURE 3.8 Reliability index for independent random variables.

two variables, following a normal distribution. For two independent random variables, following a normal distribution, a linear limit state function is given by the following relationship:

$$\beta = \frac{\mu_R - \mu_S}{\sqrt{\sigma_R^2 + \sigma_S^2}} \tag{3.10}$$

where μ_R is the mean of the yield stress (= 630.8 N/mm^2) (Sadowski et al., 2015), μ_S is the mean of the stress process (N/mm^2), σ_R is the standard deviation of the yield strength (= 33.7 N/mm^2), and σ_S is the standard deviation of the stress process (N/mm^2). The probability of failure is given by the inverse of the cumulative distribution function (Φ) of a standard normal distribution for a given β. The following relationship holds good:

$$P_f = \Phi^{-1}(-\beta) \tag{3.11}$$

Environmental loads are dynamic in effect. Buoyant legs undergo variable submergence, which results in the dynamic tension variation in the tethers. Among a group of tethers in each buoyant leg, only one is considered for the fatigue analysis. As tethers are carefully placed in a group, this assumption holds good. Figure 3.9 shows the response of the tethers of buoyant leg-1 in sea state (S_1–0°) with a maximum and minimum pretension of 35.9 MN and 20.9 MN, respectively. After performing the rainflow counting the matrix histogram, showing the stress ranges, is plotted in Figure 3.10. As seen in the figure, the cycle range is varying from 0–70 N/mm^2, but is concentrated in the region where the cycle range is less than 50 N/mm^2; the cycle average is changing in the range of 128–140 N/mm^2. Figure 3.11 shows the tether tension spectrum. The total area under the spectrum, which represents the energy, is about 4.21×10^{12} N^2-rad. Peaks are observed at the heave and pitch natural frequency of the platform, indicating a strong coupling between them.

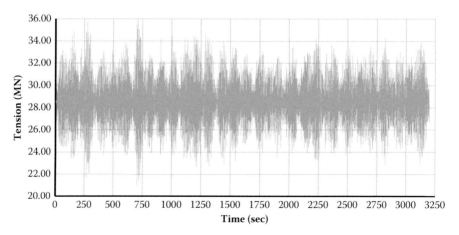

FIGURE 3.9   Dynamic tension variation of the tethers under buoyant leg-1 (S_1–0°).

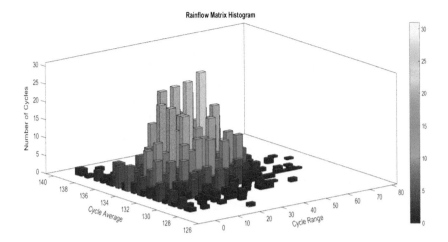

FIGURE 3.10 Rainflow matrix histogram of tethers under buoyant leg-1 (S_1–0°).

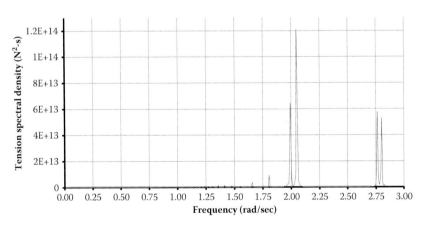

FIGURE 3.11 PSD of dynamic tension variation of tethers under buoyant leg-1 (S_1–0°).

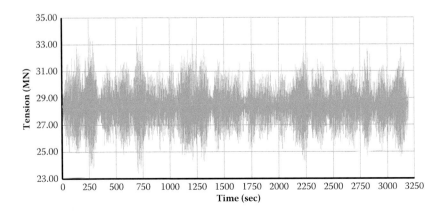

FIGURE 3.12 Dynamic tension variation of the tethers under buoyant leg-2 (S_1–0°).

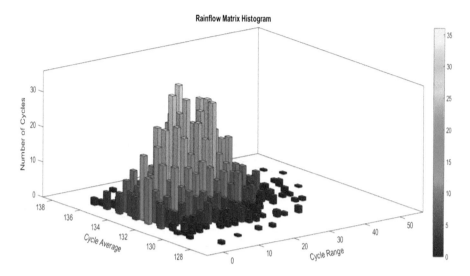

FIGURE 3.13   Rainflow matrix histogram of tethers under buoyant leg-2 (S_1–0°).

Figure 3.12 shows the response time history of tethers in buoyant leg-2 in sea state (S_1–0°) with a maximum and minimum tension of 34.6 MN and 23.4 MN, respectively. After performing the rainflow counting, the results are plotted in Figure 3.13. It is seen from the plots that the cycle range is varying from 0–50 N/mm², but the histograms are more concentrated in the region where the cycle range is lesser than 30 N/mm². The cycle average is varying in the range (128–136 N/mm²). Figure 3.14 shows the spectral plots of the tension variation. The area under the curve, which is the spectral energy of tension variations, is about 1.93×10^{12} N²-rad. It is about 45% of that of buoyant leg-1.

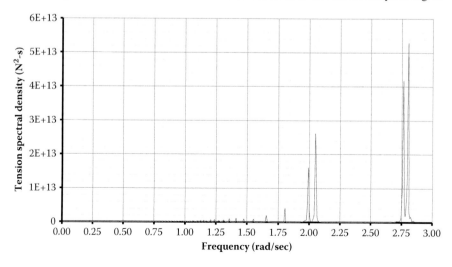

FIGURE 3.14   PSD of dynamic tension variation of tethers under buoyant leg-2 (S_1–0°).

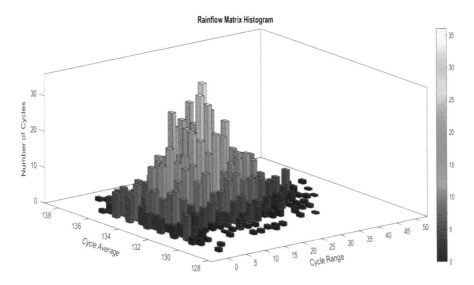

FIGURE 3.15 Rainflow matrix histogram of tethers under buoyant leg-1 (S_1–30°).

Hence, there is a relative change in the fatigue of tether-4 with respect to that of tether-1 for the same sea state.

Figure 3.15 shows the rainflow counting matrix, and Figure 3.16 shows the PSD plots of the tether tension variation for buoyant leg-1 at 30° wave-heading. It is seen from the figure that the cycle ranges vary from 0–45 N/mm², while the stress averages vary from 128–138 N/mm². The area under the PSD is about 2.24×10^{12} N²-rad, which is about 53% of that of the buoyant leg-1 at 0° wave-heading.

Figures 3.17 and 3.18 show the rainflow counting matrix PSD of the tension variation in the tether of buoyant leg-2 at a wave-heading angle of 30°, respectively.

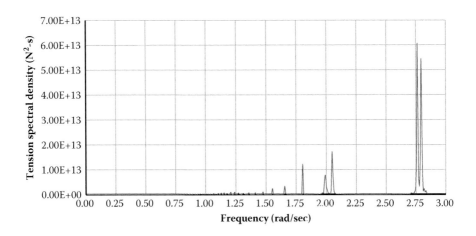

FIGURE 3.16 PSD of dynamic tension variation of tethers under buoyant leg-1 (S_1–30°).

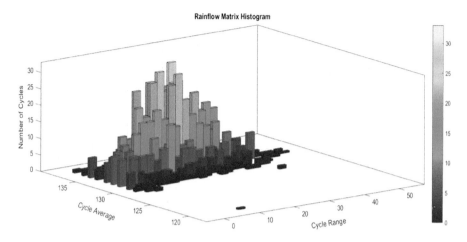

FIGURE 3.17   Rainflow matrix histogram of tethers of buoyant leg-2 (S_1–30°).

It is observed from the figures that the stress cycle averages are varying in the range (127–138 N/mm²) while the cycle ranges are varying in the range (0–50 N/mm²). The area under the tension spectrum is about 2.44 × 10¹² N²-rad, which is about 58% of that of buoyant leg-1 in the same sea state and 0° wave-heading. Figures 3.19 and 3.20 show the rainflow counting matrix and PSD of the tether tension variation in buoyant leg-3 at a wave-heading of 30°, respectively. It is seen from the figure that the cycle averages are varying in the range (128–137 N/mm²) while the cycle ranges are varying between (0–45 N/mm²). The area under the tension spectrum is computed as 2.06 × 10¹² N²-rad, which is about 49% of that of buoyant leg-1 in the same sea state at 0° wave-heading.

Figures 3.21 and 3.22 show the rainflow counting matrix and PSD of the tether tension variation in buoyant leg-1 at a 90° wave-heading, respectively. It is

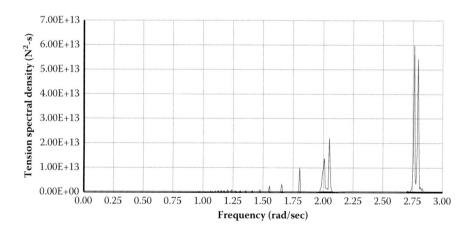

FIGURE 3.18   PSD of dynamic tension variation of tethers in buoyant leg-2 (S_1–30°).

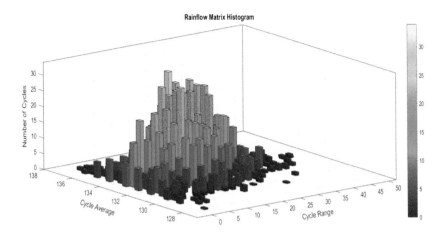

FIGURE 3.19 Rainflow matrix histogram of tethers in buoyant leg-3 (S_1–30°).

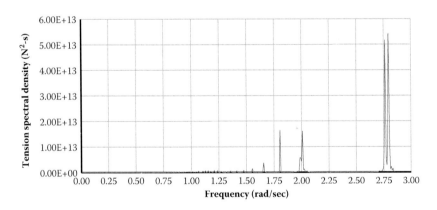

FIGURE 3.20 PSD of dynamic tension variation of tethers in buoyant leg-3 (S_1–30°).

observed from the figures that the cycle ranges vary in the range (0–50 N/mm^2), and the cycle averages vary in the range (128–136 N/mm^2). The area under the tension spectrum is 2.2795 × 10^{12} N^2-rad, which is about 54% of that of buoyant leg-1 in the same sea state with a 0° wave-heading. Figures 3.23 and 3.24 show the rainflow counting matrix and PSD of the tether tension variation in buoyant leg-2 at a wave incidence angle of 90°, respectively. It is observed that the cycle ranges are varying in the range of (0–55 N/mm^2), and the cycle averages are in the range (128–137 N/mm^2). The area under the tension spectrum is 2.8448 × 10^{12} N^2-rad, which is about 68% of that of buoyant leg-1 in the same sea state with a 0° wave-heading.

Figures 3.25 and 3.26 show the rainflow counting matrix and PSD of the tether tension variation in buoyant leg-3 for a 90° wave-heading, respectively. It is observed that the cycle ranges are between (0–50 N/mm^2), and the cycle averages are varying in the range (128–137 N/mm^2). The area under the tension spectrum is

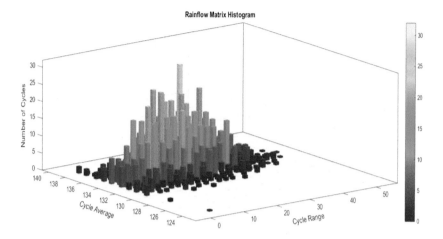

FIGURE 3.21   Rainflow matrix histogram of tethers in buoyant leg-1 (S_1–90°).

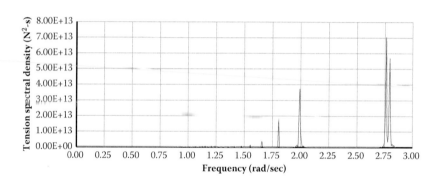

FIGURE 3.22   PSD of dynamic tension variation of tethers in buoyant leg-1 (S_1–90°).

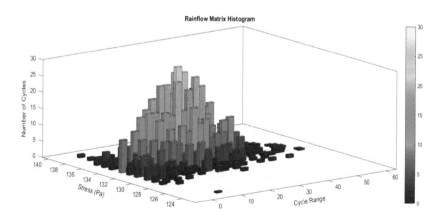

FIGURE 3.23   Rainflow matrix histogram of tethers in buoyant leg-2 (S_1–90°).

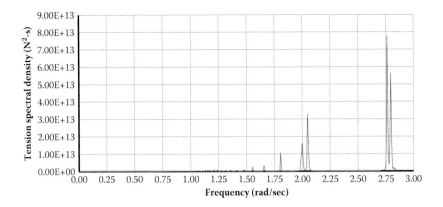

FIGURE 3.24   PSD of dynamic tension variation of tethers in buoyant leg-2 (S_1–90°).

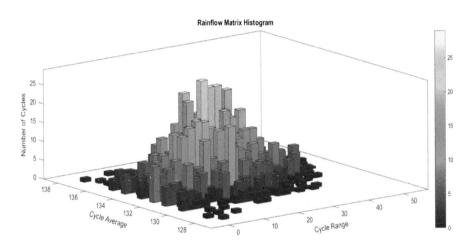

FIGURE 3.25   Rainflow matrix histogram of tethers in buoyant leg-3 (S_1–90°).

2.8432 × 10^{12} N^2-rad, which is about 68% of that of buoyant leg-1 under the same sea state with a 0° wave-heading. Figures 3.27 and 3.28 show the rainflow counting matrix and PSD of the tether tension variation in buoyant leg-1 for 150° wave-heading. It is observed that the cycle ranges are between (0–50 N/mm²), and the cycle averages are varying in the range (128–136 N/mm²). The area under the tension spectrum is 2.41 × 10^{12} N^2-rad, which is about 57% of that of buoyant leg-1 under the same sea state with a 0° wave-heading.

Figures 3.29 and 3.30 show the rainflow counting matrix and PSD of the tether tension variation in buoyant leg-2 for a 150° wave-heading, respectively. It is observed that the cycle ranges are between (0–45 N/mm²), and the cycle averages

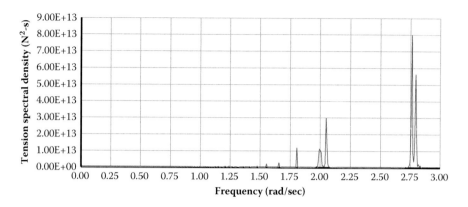

FIGURE 3.26   PSD of dynamic tension variation of tethers in buoyant leg-3 (S_1–90°).

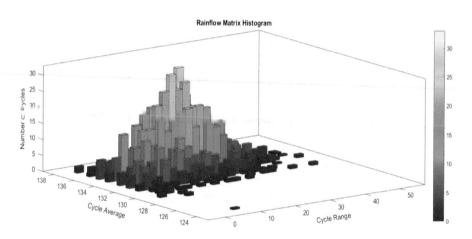

FIGURE 3.27   Rainflow matrix histogram of tethers in buoyant leg-1 (S_1–150°).

are varying in the range (128–137 N/mm²). The area under the tension spectrum is 2.12×10^{12} N²-rad, which is about 51% of that of the buoyant leg-1 under the same sea state with a 0° wave-heading. Figures 3.31 and 3.32 show the rainflow counting matrix and PSD of the tether tension variation in buoyant leg-3 for 150° wave-heading. It is observed that the cycle ranges are between (0–50 N/mm²), and the cycle averages are varying in the range (128–137 N/mm²). The area under the tension spectrum is 2.49×10^{12} N²-rad, which is about 59% of that of the buoyant leg-1 under the same sea state with a 0° wave-heading.

Figures 3.33 and 3.34 show the rainflow counting matrix and PSD of the tether tension variation in buoyant leg-1 for a 180° wave-heading, respectively. It is observed that the cycle ranges are between (0–70 N/mm²), and the cycle averages are varying in the range (127–140 N/mm²). The area under the tension spectrum

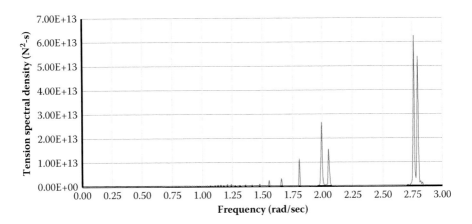

FIGURE 3.28   PSD of dynamic tension variation of tethers in buoyant leg-1 (S_1–150°).

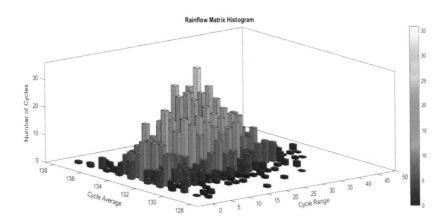

FIGURE 3.29   Rainflow matrix histogram of tethers in buoyant leg-2 (S_1–150°).

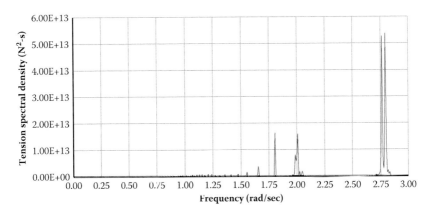

FIGURE 3.30   PSD of dynamic tension variation of tethers in buoyant leg-2 (S_1–150°).

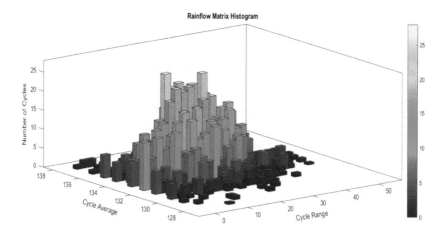

FIGURE 3.31   Rainflow matrix histogram of tethers in buoyant leg-3 (S_1–150°).

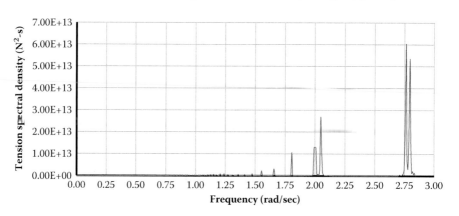

FIGURE 3.32   PSD of dynamic tension variation of tethers in buoyant leg-3 (S_1–150°).

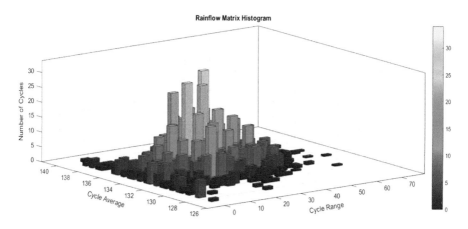

FIGURE 3.33   Rainflow matrix histogram of tethers in buoyant leg-1 (S_1–180°).

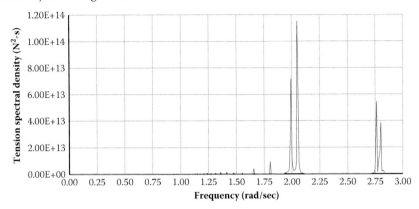

FIGURE 3.34 PSD of dynamic tension variation of tethers in buoyant leg-1 (S_1–180°).

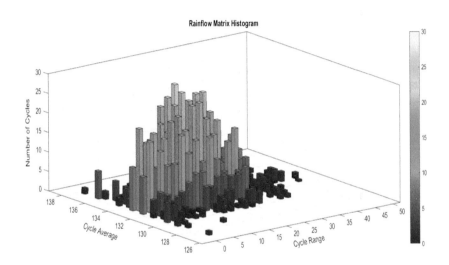

FIGURE 3.35 Rainflow matrix histogram of tethers in buoyant leg-2 (S_1–180°).

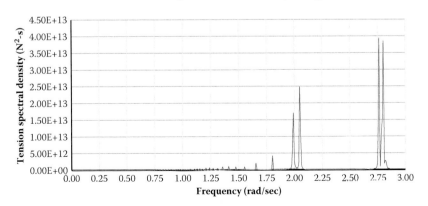

FIGURE 3.36 PSD of dynamic tension variation of tethers in buoyant leg-2 (S_1–180°).

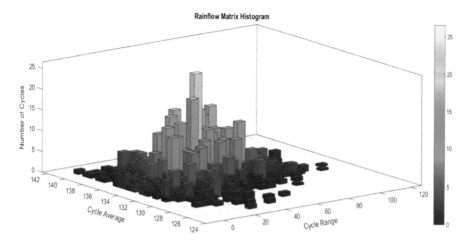

FIGURE 3.37　Rainflow matrix histogram of tethers in buoyant leg-1 ($S_4 + C_3$–0°).

is 4.39×10^{12} N^2-rad, which is about double that of buoyant leg-1 under the same sea state with a 0° wave-heading. Figures 3.35 and 3.36 show the rainflow counting matrix and PSD of the tether tension variation in buoyant leg-2 for 180° wave-heading, respectively. It is observed that the cycle ranges are between (0–45 N/mm^2), and the cycle averages are varying in the range (129–136 N/mm^2). The area under the tension spectrum is 1.98×10^{12} N^2 rad, which is about 47% of that of buoyant leg-1 under the same sea state with a 0° wave-heading.

Figures 3.37 and 3.38 show the rainflow counting matrix and PSD of the tether tension variation in buoyant leg-1 for the (S_4–0°) wave-heading in the presence of current (C_3), respectively. It is observed that the cycle ranges are between (0–100 N/mm^2), and the cycle averages vary in the range (125–140 N/mm^2). The

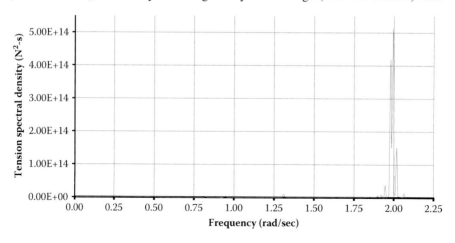

FIGURE 3.38　PSD of dynamic tension variation of tethers in buoyant leg-1 ($S_4 + C_3$–0°).

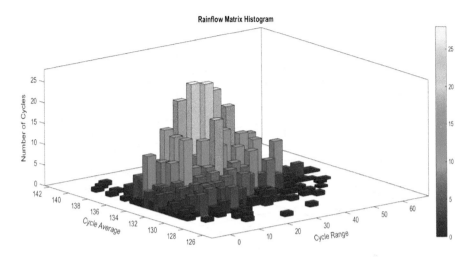

FIGURE 3.39   Rainflow matrix histogram of tethers in buoyant leg-2 ($S_4 + C_3$–0°).

area under the tension spectrum is 11.7×10^{12} N^2-rad, which is about 2.8 times that of buoyant leg-1 under the (S_1–0°) wave-heading, in the absence of current. It can be seen that the current has a significant impact on the dynamic tether tension variation. Figures 3.39 and 3.40 show the rainflow counting matrix and PSD of the tether tension variation in buoyant leg-2 for (S_4–0°) in the presence of current (C_3), respectively. It is observed that the cycle ranges are between (0–60 N/mm^2), and the cycle averages vary in the range (127–140 N/mm^2). The area under the tension spectrum is 3.37×10^{12} N^2-rad, which is about 80% of that of buoyant leg-1 under (S_1–0°), in the absence of current.

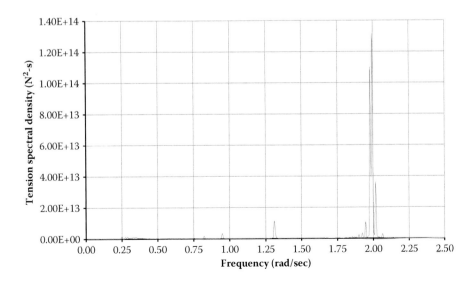

FIGURE 3.40   PSD of dynamic tension variation of tethers in buoyant leg-2 ($S_4 + C_3$–0°).

TABLE 3.7
Area under tether tension spectrum (10^{12}, N^2-rad)

Wave conditions	Leg-1	Leg-2	Leg-3
S_1-$0°$	4.210	1.930	1.897
S_1-$30°$	2.245	2.440	2.060
S_1-$90°$	2.279	2.845	2.843
S_1-$150°$	2.418	2.120	2.490
S_1-$180°$	4.391	1.980	1.960
S_1+C_1-$0°$	4.850	1.980	1.965
S_2-$0°$	4.600	1.680	1.649
S_2+C_2-$0°$	6.030	3.250	3.204
S_3-$0°$	4.720	1.490	1.538
S_3+C_3-$0°$	4.450	1.780	1.750
S_4-$0°$	3.780	1.320	1.298
S_4+C_3-$0°$	11.700	3.370	3.363

TABLE 3.8
Reliability index against yielding for Triceratops under random waves

Description	Parameters	S_1-$0°$	S_1-$30°$	S_1-$90°$	S_1-$150°$	S_1-$180°$	S_1+C_1-$0°$
Tethers under buoyant leg-1	20-year damage	0.6690	0.1251	0.219	0.1586	0.7743	1.0
	Total fatigue life (years)	29	159.87	91.32	126.10	25.82	18.46
	Reliability Index (β_{yield})	14.18	14.43	14.38	14.41	14.15	14.09
Tethers under buoyant leg-2	20-year damage	0.0667	0.1668	0.3025	0.0621	0.0859	0.0488
	Total fatigue life (years)	299.85	119.09	66.115	322.06	232.82	409.83
	Reliability Index (β_{yield})	14.50	14.43	14.37	14.47	14.49	14.49
Tethers under buoyant leg 3	20-year damage	0.0656	0.0846	0.3122	0.1593	0.0709	0.0396
	Total fatigue life (years)	304.87	236.4	64.06	125.49	282.08	505.05
	Reliability Index (β_{yield})	14.5	14.8	14.38	14.42	14.49	14.49
Description	**Parameters**	S_2-$0°$	S_2+C_2-$0°$	S_3-$0°$	S_3+C_3-$0°$	S_4-$0°$	S_4+C_3-$0°$
Tethers under buoyant leg 1	20-year damage	1.0	1.0	1.0	0.8931	1.0	1.0
	Total fatigue life (Yrs)	16.02	8.32	9.81	22.3	13.02	1.008
	Reliability Index (β_{yield})	14.13	13.95	14.09	14.15	14.23	13.30

(*Continued*)

TABLE 3.8 (Continued)

Description	Parameters	S_1–0°	S_1–30°	S_1–90°	S_1–150°	S_1–180°	S_1+C_1–0°
Tethers under	20-year damage	0.0124	0.3347	0.004	0.0378	0.0046	0.3886
buoyant leg-2	Total fatigue life (Yrs)	∞	59.755	∞	529.1	∞	51.46
	Reliability Index (β_{yield})	14.53	14.32	14.56	14.51	14.58	14.30
Tethers under	20-year damage	0.0126	0.3425	0.0026	0.0428	0.0039	0.4889
buoyant leg-3	Total fatigue life (Yrs)	∞	58.394	∞	467.2	∞	40.90
	Reliability Index (β_{yield})	14.54	14.33	14.55	14.51	14.59	14.30

In all the above spectral density plots of tension variation in tethers, it is common to note that the peaks were observed in the vicinity of either heave or pitch (or both) natural frequencies. It is also said that the dynamic tension variation is mainly due to these two movements of the platform.

3.6.3 Fatigue Life Assessment

The area under the spectral density plots of tension variation of tethers is computed for all the above cases and summarized in Table 3.7. Based on the data, the reliability index is calculated against the yielding of tethers. It is to be noted that the platform is subjected to different sea states in random waves under different wave-heading angles, as summarized earlier (please, refer to Table 3.2). The reliability index for different wave conditions is computed for each buoyant leg, namely, 1, 2, and 3. Fatigue damage is assessed under two conditions, namely, 20-year damage and total fatigue damage. It is because 20 years is usually considered as the service life of an offshore platform, beyond which the oil and gas exploration at any particular drilling site becomes unfeasible (Table 3.8).

Nonlinear regression analysis is performed to map the fatigue life and the reliability index against yielding. It is computed to estimate the damage that occurred to the tethers in the presence of a hurricane, which can be obtained by measuring the mean and

TABLE 3.9
Relationship between fatigue life and reliability index against yielding

S. No	β_{yield}	A (years)	B (no units)
1	$13.29 < \beta_{yield} < 14.379$	3.519×10^{-33}	5.496
2	$14.41 < \beta_{yield} < 14.5017$	1.095×10^{-65}	10.71
3	$\beta_{yield} > 14.5095$	4.471×10^{-153}	24.63

TABLE 3.10

Properties under the postulated failure of tethers

S. No	Properties	Before failure of 3 tethers	After failure of 3 tethers
1	Total weight of Triceratops (MN)	565.5	518.59
2	Stiffness of Restraining System (MN/m)	447.3	298.2
3	Pre-Tension per Tether (MN)	28.8	49.6

standard deviation of the stress process by installing sensors at a few critical sections on the buoyant legs and the deck. The following expression holds good:

$$L = A \ e^{(B \ x \ \beta_{yield})} \qquad (3.12)$$

where L is the fatigue life (in years), and the constants (A, B), as obtained from the regression analysis, are as listed in Table 3.9.

3.7 RELIABILITY AND FATIGUE ASSESSMENT UNDER POSTULATED FAILURE

Under normal operating conditions where tethers are not failures, the fatigue life of tethers is estimated as discussed above. Results from the above set of studies confirm that Triceratops does not fail under normal operating conditions in all the sea states under

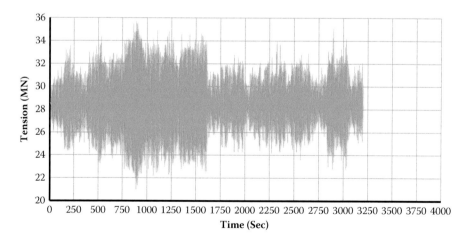

FIGURE 3.41 Tension variation of tethers of buoyant leg-1 under normal operation (S_2–0°).

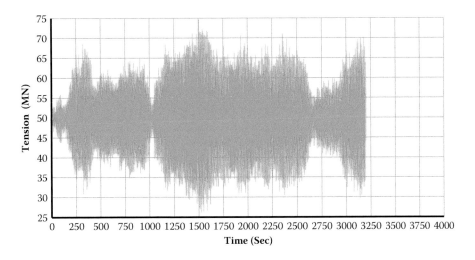

FIGURE 3.42 Tension variation of tethers of buoyant leg-1 under postulated failure (S_2–$0°$).

different wave-heading angles. Based on the tension variation in the tethers, those that have undergone the maximum difference are hypothetically disconnected, and a postulated failure condition is intuited. Under the postulated failure conditions, a marginal change in the natural periods of various degrees-of-freedom, both in the horizontal and vertical planes, is observed. Table 3.10 shows the platform's structural properties under the postulated failure conditions and compared with the normal operating conditions. It is due to the change in the mass of the platform (about 7%) and a reduction in the stiffness of the tethers (about 33%).

Figures 3.41 and 3.42 show the dynamic tension variation in one of the tethers in buoyant leg-1, before and after the postulated failure, respectively. It is observed that the mean pre-tension, which is about 28 MN, increased to about 49 MN after the

TABLE 3.11
Statistics of stress process of tethers in leg-1 under postulated failure conditions

S. No	Wave condition	Mean	Median	Mode	Standard deviation	Skewness (10^{-3})	Kurtosis (10^{-1})
1	S_2	233.7	233.6	233.7	34.9	−2.3	−3.57
2	S_2+C_2	233.7	233.8	230.3	31.5	−6.3	−5.14
3	S_3	233.7	233.7	233.7	42.5	−0.19	−9.40
4	S_3+C_3	233.8	233.6	233.8	39.6	11.14	−2.34
5	S_4	233.6	233.6	233.7	34.9	−0.86	−3.10
6	S_4+C_3	233.7	233.7	233.8	45.8	−0.51	−4.46

TABLE 3.12

Statistics of stress process of tethers in leg-2 under postulated failure conditions

S. No	Wave condition	Mean	Median	Mode	Standard deviation	Skewness (10^{-3})	Kurtosis (10^{-1})
1	S_2	232.6	232.6	226.4	24.7	−4.4	−5.3
2	S_2+C_2	232.6	232.6	240.9	25.6	−1.6	−7.9
3	S_3	232.5	232.4	232.3	22.5	−0.77	−7.2
4	S_3+C_3	232.6	232.5	232.5	28	4.2	−4.4
5	S_4	232.5	232.4	237.3	18.4	24.9	−2.4
6	S_4+C_3	232.5	232	233	23.8	29.5	−3.5

TABLE 3.13

Statistics of stress process of tethers in leg-3 under postulated failure conditions

S. No	Wave condition	Mean	Median	Mode	Standard deviation	Skewness (10^{-3})	Kurtosis (10^{-1})
1	S_2	232.6	232.6	232.6	24.7	−4.4	−5.4
2	S_2+C_2	232.6	232.6	232.6	25.9	−1.6	−8.0
3	S_3	232.5	232.5	232.6	22.3	1.0	7.7
4	S_3+C_3	232.6	232.5	232.5	27.5	−4.6	−4.4
5	S_4	232.5	232.4	232.6	18.1	−25.6	−2.3
6	S_4+C_3	232.5	232.4	240.6	23.5	−30.6	−3.9

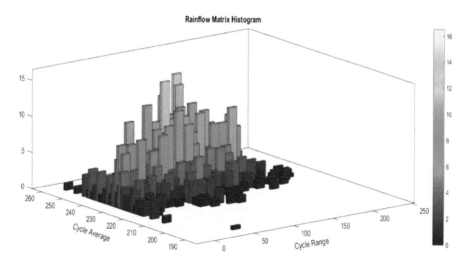

FIGURE 3.43 Rainflow histogram of tethers in buoyant leg-1 under postulated failure (S_2–0°).

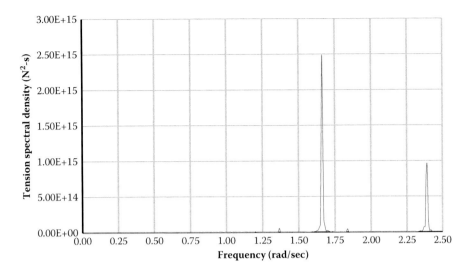

FIGURE 3.44   PSD of tension variation in buoyant leg-1 under postulated failure (S_2–0°).

postulated failure. A significant increase in the mean pre-tension, as observed, resulted in the high-cycle ranges and averages. It subsequently reduced the fatigue life of the tethers under buoyant leg-1 to 0.03 years, with a reliability index of 8.18 for yielding; it has a probability of failure of $1.149 \times 10\text{–}16$. Hence, postulated failure reduces the fatigue life of the platform considerably. It is also influenced by the wave approach angle, significantly. Postulated failure induced in tethers of each buoyant leg shifts the axial pre-tension to a higher magnitude. Fatigue life reduces to a highly

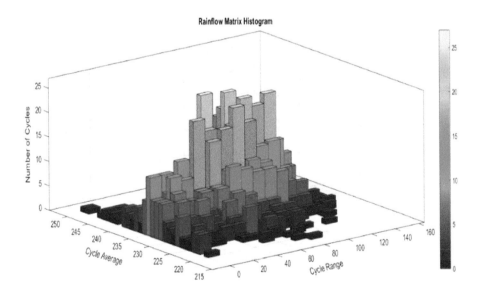

FIGURE 3.45   Rainflow histogram of tethers in buoyant leg-2 under postulated failure (S_2–0°).

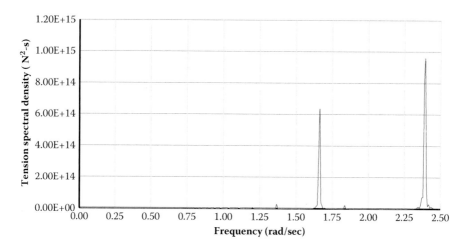

FIGURE 3.46   PSD of tension variation in buoyant leg-2 under postulated failure (S_2–0°).

alarming value. Fatigue life resulting from the dynamic tether tension variation is a function of the mean and standard deviation of the axial stress variations in tethers. Heave, even though being one of the stiff degrees-of-freedom, is significantly influenced by the postulated failure conditions. It induces dynamic tether tension variations, resulting in a phase lag in the heave motion of buoyant legs. Higher values of reliability indices for yielding indicate that failure due to yielding is almost impossible even under the postulated failure conditions. Tables 3.11, 3.12, and 3.13

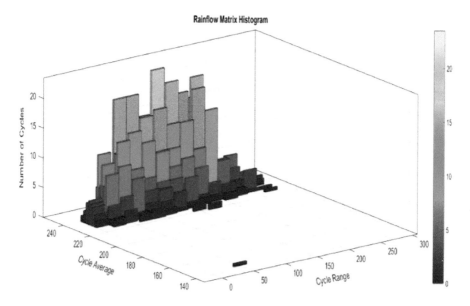

FIGURE 3.47   Rainflow histogram of tethers in buoyant leg-1 under postulated failure (S_4 – C_3–0°).

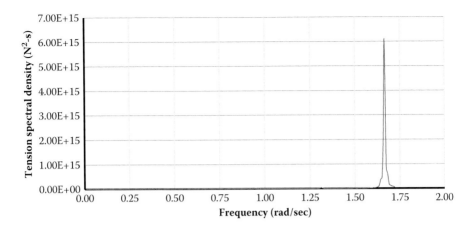

FIGURE 3.48 PSD of tension variation in buoyant leg-1 under postulated failure $(S_4 - C_3 - 0°)$.

summarize the statistical values of the tension variations of tethers in all the buoyant legs, namely 1, 2, and 3, respectively. It is observed from the tables that the mean, median, and mode are equal in all three legs. The skewness and kurtosis are zero, confirming that the stress process is symmetric even under the postulated failure condition. Hence, a normal distribution is assumed.

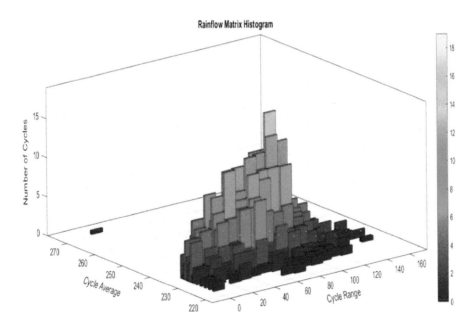

FIGURE 3.49 Rainflow histogram of tethers in buoyant leg-2 under postulated failure $(S_4 - C_3 - 0°)$.

The rainflow counting matrix of the stress cycle for buoyant leg-1 is plotted in Figure 3.43 for (S_2–0°). It can be seen that the cycle ranges vary between (0–220 N/mm^2) with cycle averages varying in the range (210–250 N/mm^2). The power spectral density plot of the dynamic tension variations is shown in Figure 3.44. The area under the plot is 5.4866 × 10^{13} N^2-rad, which is about 13 times higher than that of the tension spectrum of the buoyant leg-1 for (S_1–0°) under no-failure conditions. Because of the high cycle ranges and averages, the fatigue life of the tethers in buoyant leg-1 is 0.03 years and the reliability index is 8.18 for yielding; it has a probability of failure of 1.149 × 10^{-16}.

The rainflow counting matrix of stress cycle for buoyant leg-2 is plotted in Figure 3.45 for (S_2–0°). It can be seen that the cycle ranges vary between (0–140 N/mm^2) with cycle averages varying in the range (220–250 N/mm^2). The power spectral density plot of the dynamic tension variations is shown in Figure 3.46. The area under the plot is 2.767 × 10^{13} N^2-rad, which is about seven times higher than that of the tension spectrum of the buoyant leg-1 for (S_1–0°) under

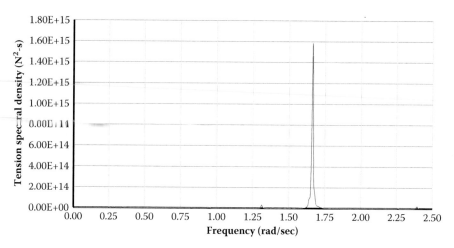

FIGURE 3.50 PSD of tension variation in buoyant leg-2 under postulated failure ($S_4 - C_3 - 0°$).

TABLE 3.14

Fatigue life and reliability index under postulated failure conditions

Description	Parameters	S_2-0°	S_2+C_2-0°	S_3-0°	S_3+C_3-0°	S_4-0°	S_4+C_3-0°
Tethers under buoyant leg-1	Fatigue life (Yr)	0.03	0.04	0.02	0.02	0.03	0.01
	β_{yield}	8.18	8.59	7.31	7.63	8.18	6.97
Tethers under buoyant leg-2	Fatigue life (Yr)	0.09	0.07	0.18	0.05	0.34	0.12
	β_{yield}	9.51	9.39	9.81	9.08	10.3	9.64
Tethers under buoyant leg-3	Fatigue life (Yr)	0.09	0.07	0.20	0.06	0.36	0.12
	β_{yield}	9.52	9.40	9.84	9.14	10.4	9.69

TABLE 3.15

Area under tether tension spectrum (10^{13} N^2-rad)

Parameters	S_2-0°	S_2+C_2-0°	S_3-0°	S_3+C_3-0°	S_4-0°	S_4+C_3-0°
Tethers in buoyant leg-1	5.487	4.463	8.186	7.073	5.539	9.248
Tethers in buoyant leg-2	2.767	2.922	2.31	3.54	1.54	2.57
Tethers in buoyant leg-3	2.76	2.902	2.27	3.42	1.5	2.5

no-failure conditions. Because of the high cycle ranges and averages, the fatigue life of the tethers in buoyant leg-1 is 0.09 years and the reliability index is 10.3 for yielding; it has a probability of failure of 3.523×10^{-25}.

The presence of current creates further vulnerability to the platform under the postulated failure conditions. In the sea state (S_4+C_3-0°), cycle ranges and cycle averages are observed to be higher, as seen in Figure 3.47. It is observed that the average cycle stress is mostly concentrated at 230 N/mm^2, and the cycle ranges vary in the range (0–270 N/mm^2). Figure 3.48 shows the spectral density plots under this case of failure, in the presence of current. It is seen that the area under the spectral plot is about 9.248×10^{13} N^2-rad, which is about 22 times higher than that of buoyant leg-1 at 0° wave-heading without postulated failure. The fatigue life of tethers in the buoyant leg-1 is about 0.01 years, as the stress ranges are very high. The fatigue life is very low in this sea state, with a reliability index of 6.97 at yielding; it has a probability of failure of 1.585×10^{-12}.

In the sea state (S_4+C_3-0°), Figure 3.49 shows the rainflow counting histogram for tethers in buoyant leg-2 under postulated failure. It is observed that the cycle ranges and cycle averages are higher. The average cycle stress is mostly concentrated at 230 N/mm^2, and the cycle ranges vary in the range (0–150 N/mm^2). Figure 3.50 shows the spectral density plots under this case of failure, in the

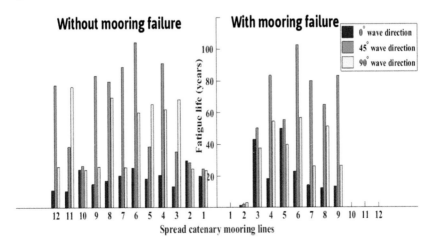

FIGURE 3.51 Fatigue life of catenary mooring lines for 10 years return period.

presence of current. It is seen that the area under the spectral plot is about 2.57 × 10^{13} N^2-rad, which is about six times higher than that of buoyant leg-1 at a 0° wave-heading without postulated failure. The fatigue life of tethers in buoyant leg-2 is about 0.12 with the reliability index of 9.64 at yielding; it has a probability of failure of 2.709 × 10^{-22}. Table 3.14 shows the fatigue life and the reliability index under different sea states, while Table 3.15 shows the summary of the area under the spectral plots under the postulated failure conditions.

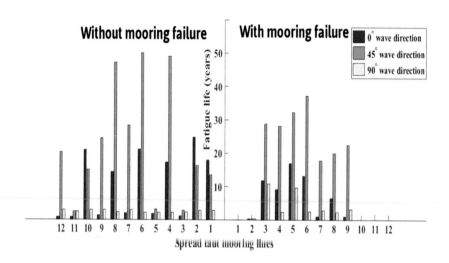

FIGURE 3.52   Fatigue life of taut mooring lines for 10-year return period.

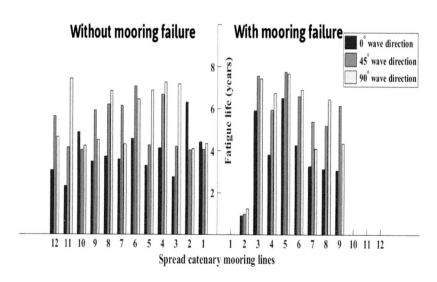

FIGURE 3.53   Fatigue life of catenary mooring lines for 100-year return period.

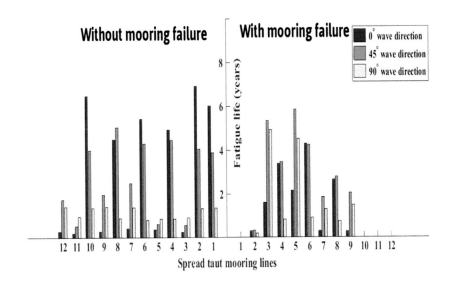

FIGURE 3.54 Fatigue life of taut mooring lines for 100-year return period.

TABLE 3.16
Statistics and fatigue life without a buoy (0°, 10 Yr)

Mooring line	Mean	Standard deviation	Fatigue life (years)
1	312.6672	15.6995	2.1677
2	310.4902	14.7625	2.5066
3	308.2175	13.9030	2.8946
4	279.4092	12.5411	4.6623
5	277.7939	12.9710	4.5195
6	276.1928	13.4125	3.9740
7	274.9681	13.8613	3.6610
8	274.8461	13.9163	3.6328
9	276.0584	13.4685	3.9417
10	277.6469	13.0270	4.2705
11	279.2504	12.5956	4.6228
12	305.6507	13.0718	3.3458
13	308.0026	13.8123	2.9243
14	310.0987	14.6516	2.5392
15	312.4705	15.5903	2.1942
16	305.8710	13.1491	3.3173

3.8 FATIGUE ANALYSIS OF MOORING LINES IN SEMI-SUBMERSIBLES

As discussed in the previous chapter, tension variation is induced in the mooring lines. A postulated failure scenario was assumed to estimate the tension variation in the mooring lines. These tension variations result in stress cycles, varying in a wide range. Hence, the fatigue life of the mooring lines, with and without a failure condition, is estimated using the Palmgren-Miner rule. The stress in the mooring

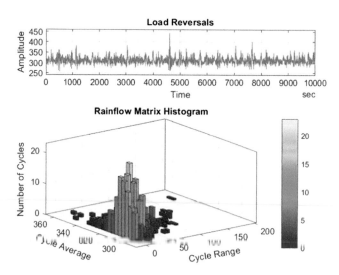

FIGURE 3.55   Mooring line #1 with minimum fatigue life without a buoy (0°, 10 Y).

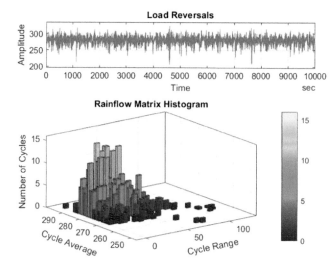

FIGURE 3.56   Mooring line #4 with maximum fatigue life without a buoy (0°, 10 Y).

TABLE 3.17
Statistics and fatigue life without a buoy (45°, 10 Yr)

Mooring line	Mean	Standard deviation	Fatigue life (years)
1	295.6560	11.8004	4.3048
2	294.3377	11.6953	4.5071
3	293.0245	11.6637	4.6176
4	279.7801	13.6788	3.0066
5	279.2702	13.7196	2.9895
6	278.7222	13.7339	3.0156
7	278.5078	13.7262	3.0462
8	287.2576	10.8121	7.4377
9	288.3260	10.6128	7.9279
10	289.7168	10.4809	8.7753
11	290.9968	10.4211	8.3250
12	304.9847	13.1941	2.9623
13	305.7501	13.3397	2.8738
14	306.2312	13.4533	2.8129
15	306.9551	13.5482	2.7769
16	291.7236	11.7002	4.6253

TABLE 3.18
Statistics and fatigue life without a buoy (90°, 10 Yr)

Mooring line	Mean	Standard deviation	Fatigue life (years)
1	285.1816	13.4643	2.3973
2	284.7077	13.6003	2.3175
3	284.2923	13.7493	2.3972
4	283.9429	13.9022	2.1519
5	284.3047	13.7499	2.2332
6	284.5895	13.6026	2.3167
7	285.1969	13.4651	2.3967
8	297.7858	11.3587	5.1756
9	298.1563	11.4149	4.8423
10	298.7988	11.5008	4.7445
11	299.2254	11.6067	4.6067
12	299.2126	11.6069	4.6067
13	298.7840	11.5012	4.7439
14	298.1398	11.4156	4.8409
15	297.7675	11.3598	4.8677
16	283.9320	13.9018	2.1523

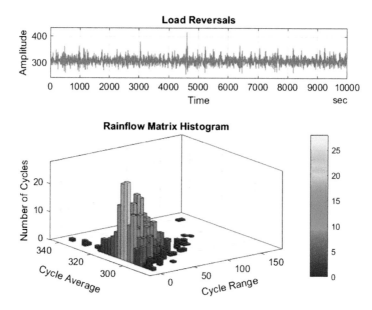

FIGURE 3.57   Mooring line #15 with minimum fatigue life without a buoy (45°, 10 Y).

lines is sufficient enough to induce fatigue failure. The S-N curve is used to estimate the number of cycles, and then the Palmgren-Miner rule is employed to estimate the fatigue life of the remaining mooring lines. The fatigue life of the mooring lines under various directions of wave, wind, and current for 10 years and 100 years return period is calculated.

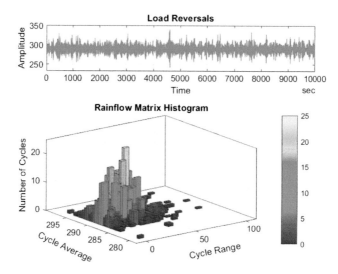

FIGURE 3.58   Mooring line #10 with maximum fatigue life without a buoy (45°, 10 Y).

3.8.1 FATIGUE LIFE OF 12-POINT MOORING

A postulate failure condition is inducted by removing mooring lines #11 and #12 of a 12-point mooring layout in catenary mooring. (For details, please see

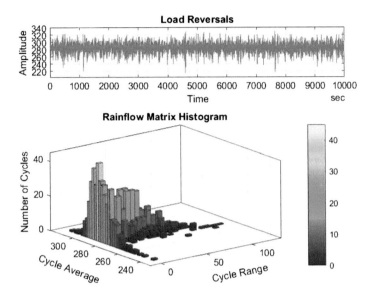

FIGURE 3.59   Mooring line #4 with minimum fatigue life without a buoy (90°, 10 Y).

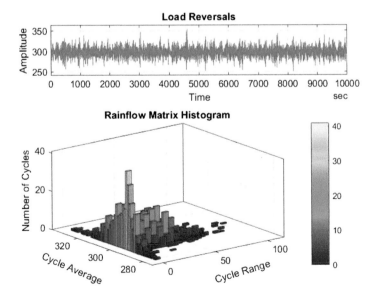

FIGURE 3.60   Mooring line #8 with maximum fatigue life without a buoy (90°, 10 Y).

TABLE 3.19

Statistics and fatigue life with a buoy (0°, 10 Yr)

Mooring line	Mean	Standard deviation	Fatigue life (years)
1	236.6933	10.6598	19.2947
2	206.6939	7.7155	57.7767
3	205.1380	8.0785	50.8445
4	203.6323	8.4334	45.2401
5	202.4426	8.7907	39.8069
6	202.3343	8.8301	39.3666
7	203.5125	8.4755	54.4725
8	205.0062	8.1224	50.1130
9	206.5503	7.7598	57.0437
10	233.8119	9.4123	19.0243
11	236.4602	10.5096	14.0595
12	238.9094	11.7054	10.4450
13	241.6126	13.0227	7.7402
14	234.0467	9.5477	18.2953
15	241.8335	13.1900	7.4857
16	239.2984	11.8854	10.0311

TABLE 3.20

Statistics and fatigue life with a buoy (45°, 10 Yr)

Mooring line	Mean	Standard deviation	Fatigue life (years)
1	220.3863	6.6283	66.4530
2	206.8614	8.4210	31.0948
3	206.3536	8.4928	30.4670
4	205.8398	8.5501	29.9415
5	205.5962	8.6072	29.3737
6	214.2194	6.1609	120.0323
7	215.3503	6.0604	128.8532
8	216.7496	6.0162	131.1120
9	218.0800	6.0249	129.5434
10	233.3205	9.2870	19.5669
11	234.1951	9.5118	18.1727
12	234.7995	9.6904	17.1934
13	235.5799	9.8528	16.3461
14	219.0237	6.6000	69.0349
15	223.1684	6.8710	55.7581
16	221.7703	6.7169	62.0330

TABLE 3.21
Statistics and fatigue life with a buoy (90°, 10 Yr)

Mooring line	Mean	Standard deviation	Fatigue life (years)
1	211.3933	9.6177	15.7939
2	211.0399	9.7387	15.1654
3	211.4028	9.6195	15.7808
4	211.7203	9.4971	16.3945
5	212.2975	9.3825	16.9065
6	225.4775	8.5127	29.6729
7	225.9413	8.6771	28.4298
8	226.6123	8.8546	26.9415
9	227.0912	9.0308	25.3987
10	227.0809	9.0320	25.3889
11	226.6005	8.8561	26.9263
12	225.9280	8.6790	28.4063
13	225.4629	8.5150	29.6141
14	211.0316	9.7370	15.1768
15	212.2859	9.3805	16.9226
16	211.8107	9.4966	16.3816

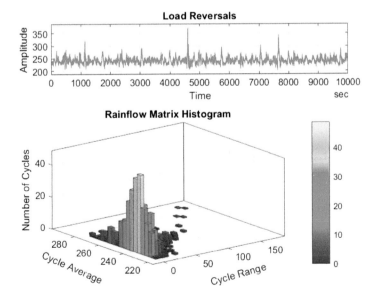

FIGURE 3.61   Mooring line #15 with minimum fatigue life with a buoy (0°, 10 Y).

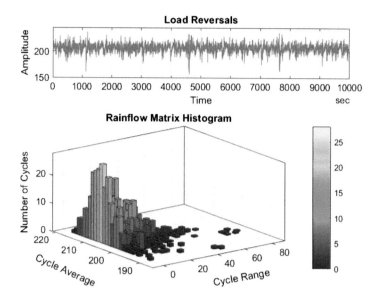

FIGURE 3.62   Mooring line #2 with maximum fatigue life with a buoy (0°, 10 Y).

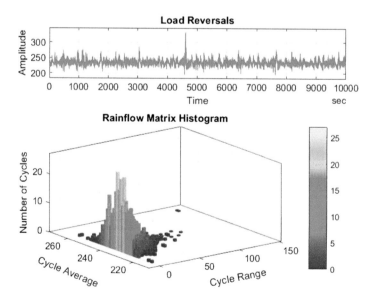

FIGURE 3.63   Mooring line #13 with minimum fatigue life with a buoy (45°, 10 Y).

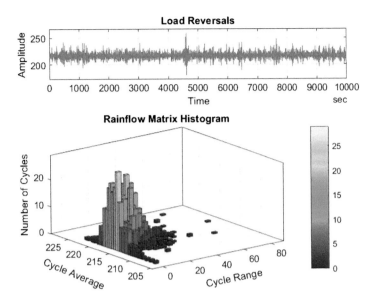

FIGURE 3.64   Mooring line #8 with maximum fatigue life with a buoy (45°, 10 Y).

Chapter 2.). The fatigue life of the mooring lines, with and without the postulated failure, is shown in Figure 3.51.

As seen in the figure, the maximum fatigue life is observed in mooring line #6 at a 45° wave-heading and is about 104.27 years, while that of the mooring line with

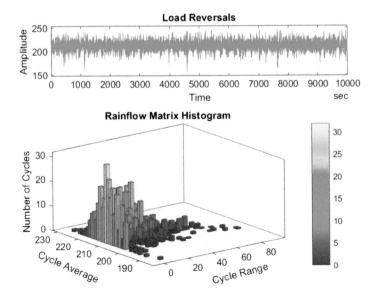

FIGURE 3.65   Mooring line #2 with minimum fatigue life with a buoy (90°, 10 Y).

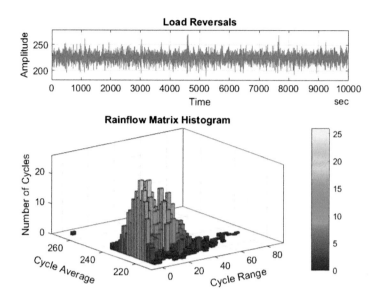

FIGURE 3.66 Mooring line #6 with maximum fatigue life with buoy (90°, 10Y).

TABLE 3.22

Statistics and fatigue life without a buoy under postulated failure (0°, 10 Yr)

Mooring line	Mean	Standard deviation	Fatigue life (years)
1	Detached	Detached	Detached
2	356.4465	19.2329	1.1860
3	346.8476	17.1629	1.5878
4	263.2097	12.6647	4.7736
5	259.8480	13.0344	4.4603
6	256.7464	13.4090	4.1593
7	254.1599	13.7787	3.8737
8	254.6528	13.8637	3.8271
9	257.2825	13.4950	4.1065
10	260.4312	13.1200	4.4032
11	263.8417	12.7479	4.7143
12	339.2926	15.2860	2.0750
13	348.6057	16.9731	1.5892
14	Detached	Detached	Detached
15	367.8890	21.3502	0.8795
16	337.5494	15.4414	2.0793

TABLE 3.23
Statistics and fatigue life without a buoy untder postulated failure (45°, 10 Yr)

Mooring line	Mean	Standard deviation	Fatigue life (years)
1	325.7931	12.5109	3.4182
2	326.7437	12.3580	3.2197
3	327.3847	12.2972	3.2799
4	Detached	Detached	Detached
5	306.3883	14.7172	2.1847
6	303.5293	14.6966	2.2371
7	301.0102	14.6458	2.3053
8	268.2010	10.6489	8.8955
9	267.6742	10.4466	9.6725
10	267.5080	10.3007	10.2996
11	267.3890	10.2134	10.6573
12	277.6592	12.3943	4.1889
13	279.4951	12.5327	4.0252
14	281.3468	12.6652	3.8756
15	283.5883	12.7977	3.7296
16	Detached	Detached	Detached

TABLE 3.24
Statistics and fatigue life without a buoy under postulated failure (90°, 10 Yr)

Mooring line	Mean	Standard deviation	Fatigue life (years)
1	285.7371	13.5888	2.3747
2	289.2831	13.6811	2.3089
3	293.0418	13.8109	2.2265
4	Detached	Detached	Detached
5	Detached	Detached	Detached
6	351.5527	13.9255	2.0113
7	353.2060	13.6394	2.1212
8	298.2891	11.2455	5.5594
9	294.0578	11.3110	5.1990
10	290.3185	11.3906	5.1403
11	286.6329	11.4786	5.0647
12	259.6943	11.6269	5.2690
13	258.8498	11.6331	5.2510
14	258.1223	11.6620	5.1826
15	257.7809	11.7146	5.0665
16	296.9958	13.9704	2.1337

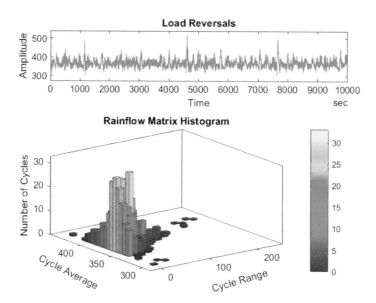

FIGURE 3.67   Mooring line #15 with minimum fatigue life without a buoy under postulated failure (0°, 10 Y).

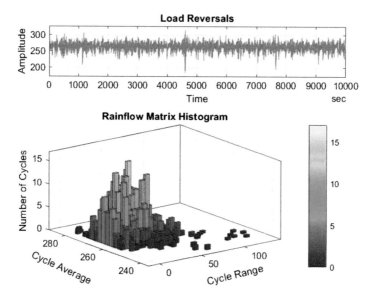

FIGURE 3.68   Mooring line #4 with maximum fatigue life without a buoy under postulated failure (0°, 10 Y).

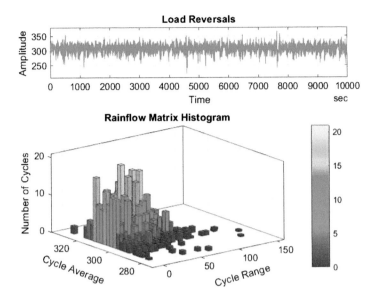

FIGURE 3.69 Mooring line #5 with minimum fatigue life without a buoy under postulated failure (45°, 10 Y).

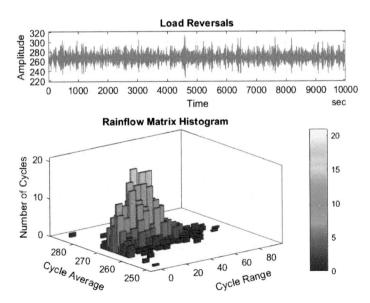

FIGURE 3.70 Mooring line #11 with maximum fatigue life without a buoy under postulated failure (45°, 10 Y).

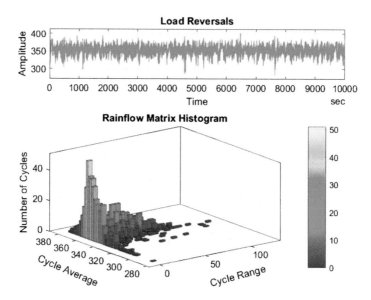

FIGURE 3.71   Mooring line #6 with minimum fatigue life without a buoy under postulated failure (90°, 10 Y).

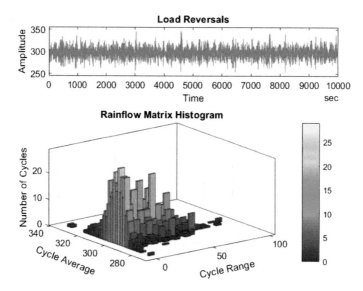

FIGURE 3.72   Mooring line #8 with maximum fatigue life without a buoy under postulated failure (90°, 10 Y).

TABLE 3.25

Statistics and fatigue life with a buoy under postulated failure (0°, 10 Yr)

Mooring line	Mean	Standard deviation	Fatigue life (years)
1	230.8337	10.4106	15.4294
2	184.9685	6.8117	117.5898
3	183.6063	7.0185	107.1538
4	182.4221	7.2309	95.9460
5	181.5665	7.4570	84.7922
6	202.1114	9.1460	33.3522
7	205.6973	8.9763	34.7178
8	209.8462	8.7918	38.5106
9	214.2997	8.5772	40.0222
10	Detached	Detached	Detached
11	316.1372	21.3998	2.6308
12	Detached	Detached	Detached
13	332.9886	28.6505	0.6773
14	224.8963	9.0586	50.4021
15	244.4190	14.0945	6.2719
16	237.3273	12.0828	9.8579

TABLE 3.26

Statistics and fatigue life with a buoy under postulated failure (45°, 10 Yr)

Mooring line	Mean	Standard deviation	Fatigue life (years)
1	203.1238	6.2280	118.5872
2	192.6675	7.8579	50.3541
3	192.9864	7.9537	60.2560
4	193.3903	8.0463	45.3153
5	194.0913	8.1449	42.5244
6	233.3529	7.4078	51.1744
7	236.7351	7.3923	72.0800
8	240.4823	7.4618	48.1325
9	Detached	Detached	Detached
10	Detached	Detached	Detached
11	266.5585	13.0414	6.3604
12	264.8302	13.0783	6.3956
13	263.1289	13.0532	6.5098
14	201.3214	6.1866	118.7338
15	207.2526	6.4586	86.6887
16	205.1012	6.3161	99.1246

TABLE 3.27

Statistics and fatigue life with a buoy under postulated failure (90°, 10 Yr)

Mooring line	Mean	Standard deviation	Fatigue life (years)
1	230.7217	10.3398	10.8097
2	Detached	Detached	Detached
3	230.7411	10.3427	10.7990
4	229.7275	10.1264	12.6698
5	228.9219	9.9246	12.2736
6	209.9735	8.3177	42.4833
7	209.1985	8.4362	41.1740
8	208.6898	8.5622	39.4596
9	208.1587	8.6838	38.1810
10	208.1477	8.6852	38.1639
11	208.6769	8.5640	39.8757
12	209.1837	8.4384	41.3836
13	209.9569	8.3204	42.3798
14	Detached	Detached	Detached
15	228.8990	9.9216	12.2873
16	229.8453	10.1260	11.6624

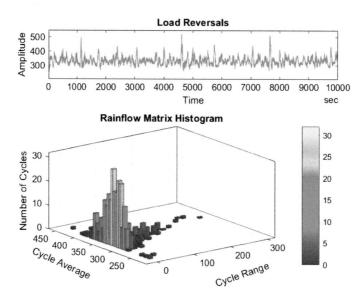

FIGURE 3.73 Mooring line #13 with minimum fatigue life with a buoy under postulated failure (0°, 10 Y).

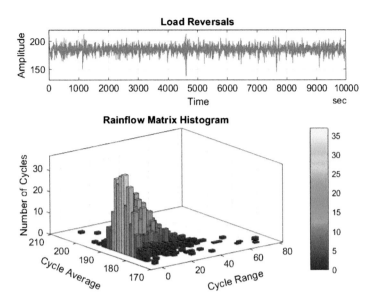

FIGURE 3.74 Mooring line #2 with maximum fatigue life with a buoy under postulated failure (0°, 10 Y).

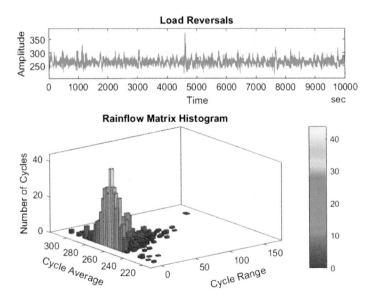

FIGURE 3.75 Mooring line #11 with minimum fatigue life with a buoy under postulated failure (45°, 10 Y).

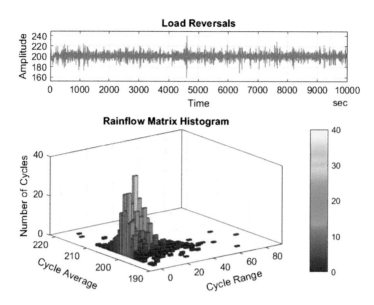

FIGURE 3.76 Mooring line #14 with maximum fatigue life with a buoy under postulated failure (45°, 10 Y).

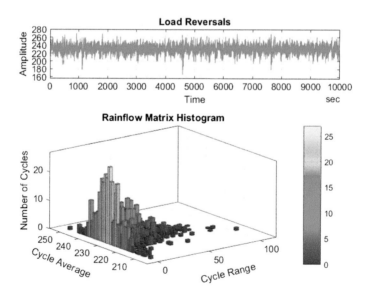

FIGURE 3.77 Mooring line #3 with minimum fatigue life with a buoy under postulated failure (90°, 10 Y).

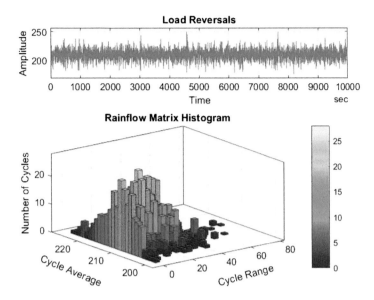

FIGURE 3.78   Mooring line #6 with maximum fatigue life with a buoy under postulated failure (90°, 10 Y).

failure is about 102.47 years. The fatigue life during a 90° wave loading is the least, and it is due to the geometric layout.

Figure 3.52 shows the fatigue life of taut-mooring lines for a ten year return period. In comparison with that of the catenary mooring, the fatigue life of the former is significantly less. It is due to the large number of stress cycles developed in the taut-mooring lines. Figures 3.53 and 3.54 show the fatigue life of all the mooring lines in both the catenary and taut-mooring under a 100 years return period. It is observed that the fatigue life for catenary mooring lines under a 100 years return period is the least for a 0° wave-heading compared to other wave-heading angles. It is due to the mooring layout to the wave-heading angles.

3.8.2 Fatigue Life of a 16-point Mooring with Submerged Buoy

As discussed in the previous chapter, the presence of a submerged buoy reduces the platform motion significantly. It offers higher damping in the heave motion while keeping the lower segment of the mooring line flexible. Massive displacements of the semi-submersible induce a significant tension variation in the mooring lines. For brevity, rainflow counting plots of the moorings with maximum and least fatigue lives are discussed. Table 3.16 summarizes the statistics of the tension variation and the fatigue life of all the mooring lines, without a buoy; 0° wave-heading and no-postulated failure conditions are considered. It is seen from the table that mooring lines #1 and #4 show the minimum and maximum fatigue life, respectively. Figures 3.55 and 3.56 show the plots of stress cycles and rainflow counting histogram of mooring lines #1 and #4, respectively.

Tables 3.17 and 3.18 show the tension variation statistics and fatigue life of all the mooring lines, without submerged a buoy; no postulated failure condition is assumed, and the wave-headings are 45° and 90°, respectively. Mooring lines with the minimum and maximum fatigue life are also highlighted.

Figures 3.57, 3.58, 3.59, and 3.60 show the plots of the stress cycles and the rainflow counting histogram for mooring lines with minimum and maximum fatigue life, for 45° and 90° wave-heading, respectively.

In the presence of the submerged buoy, tension variation in the mooring lines changes significantly. Statistics of the tension variations and the fatigue life are summarized under the no-failure condition of the platform for different wave-headings, namely, 0°, 45°, and 90°, respectively. Tables 3.19, 3.20, and 3.21, respectively, summarize the values; mooring lines with the minimum and maximum fatigue life are also highlighted. Figures 3.61, 3.62, 3.63, 3.64, 3.65, and 3.66 show the plots of stress-cycle variation and the rainflow counting histogram for the mooring lines with the minimum and maximum fatigue life under different wave-heading angles of 0°, 45°, and 90°, respectively. Submerged buoys, one in each mooring, are assumed to be deployed but not the postulated failure.

Statistics of the tension variations and the fatigue life are summarized under the postulated failure condition, in the absence of the submerged buoy in each mooring line. Results are summarized for different wave-headings, namely, at 0°, 45°, and 90°, respectively. Mooring lines with the highest stress variation are hypothetically removed to induce a postulated failure. Under such conditions, the tension in re-maining (intact) mooring lines are investigated for their fatigue life. Tables 3.22, 3.23, and 3.24 summarize the values; detached mooring lines are also highlighted. Figures 3.67, 3.68, 3.69, 3.70, 3.71, and 3.72 show the plots of stress-cycle variation and the rainflow counting histogram for the mooring lines with the minimum and maximum fatigue life under different wave-heading angles of 0°, 45°, and 90°, respectively.

Statistics of the tension variations and the fatigue life are summarized under the postulated failure condition, in the presence of the submerged buoy in each mooring line. Results are summarized for different wave-heading, namely, of 0°, 45°, and 90°, respectively. Postulated failure is induced in those mooring lines that have the highest stress variations. Under such conditions, the tensions in the remaining (intact) mooring lines are investigated for their fatigue life. Tables 3.25, 3.26, and 3.27 summarize the values; detached mooring lines are also highlighted. Figures 3.73, 3.74, 3.75, 3.76, 3.77, and 3.78 show the plots of stress-cycle variation and the rainflow counting histogram for the mooring lines with the minimum and maximum fatigue life under different wave-heading angles of 0°, 45°, and 90°, respectively.

As seen from the above case study, the presence of submerged buoys im-proved the fatigue lives of the moorings by reducing the tension in the tethers. It resulted in the reduction of the mean stress in the moorings. Further, under the postulated failure conditions, when a tether fails in the bundle of tethers, other tethers, which are in the same bundle, are likely to have less fatigue life. This is due to the load-sharing process that is triggered. Under normal circumstances, in the absence of the submerged buoy in each of the mooring lines, fatigue life is more under 45° angle of wave incidence. The average fatigue life is about 5.35 years for 45°, while for 0°, it is 3.5 years.

3.9 FATIGUE LIFE USING MATLAB CODE

In this section, we will explain the procedure to assess the fatigue life, reliability index, and the probability of failure using a computer program on a MATLAB application. One of the pre-requisites to use this program is to obtain reliable input data. As discussed in detail, the offshore platform is analyzed for different environmental loads to obtain the dynamic tension variations in the mooring lines or the tethers as the case may be. Therefore, it is now understood that the reader has the Excel file, which contains the time history of the stress variation of the mooring line, which is to be further analyzed for the fatigue life.

Step 1: Save the stress response as the Excel sheet and transfer to MATLAB.

Step 2: Extract the Vibration data zip file (it is a free-ware and is compatible to run with MATLAB). Extract it in the same folder where the MATLAB program exists.

Step 3: Open the MATLAB, and go to the current folder where the vibration data zip file is extracted.

Step 4: Now, right-click on the vibration data, and select Add to Path and Selected Folders and Subfolders, as shown in the figure.

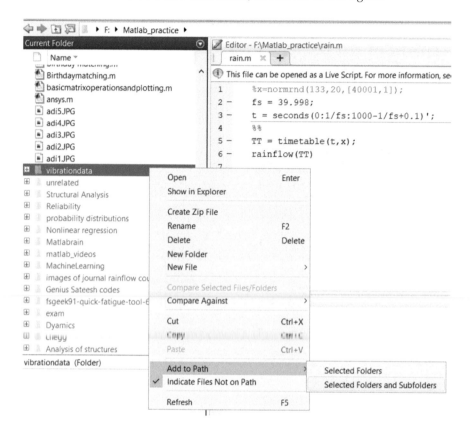

Step 5: Now, enter the vibration data on the command window. The vibration data toolbox should pop up.

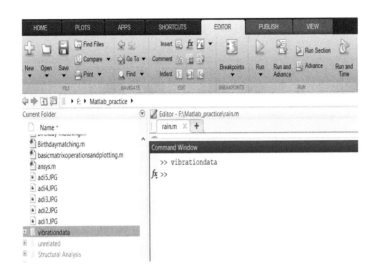

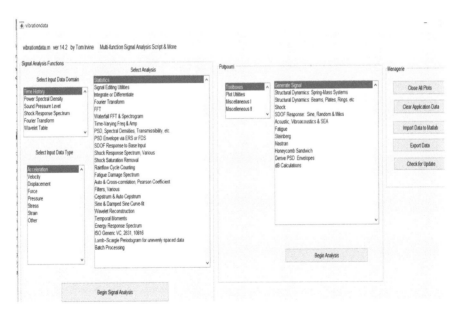

Step 6: Select the input data domain as Time History and the data type as Stress. Select the Rainflow Counting algorithm in the analysis and Fatigue in the toolboxes.

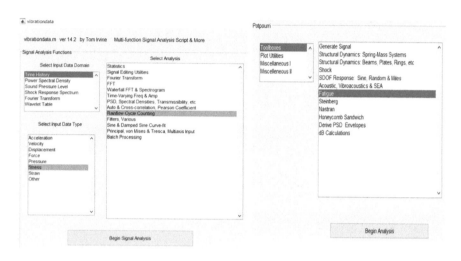

Step 7: Click on Begin the Analysis. It will pop up as a fatigue toolbox, as shown in the figure.

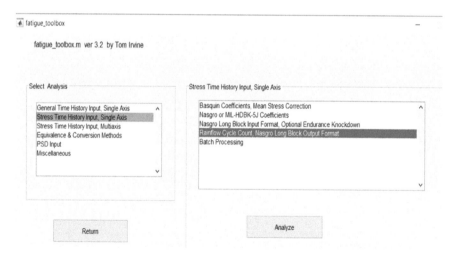

Step 8: Select the stress Time history Input, Single Axis in the analysis, and Rainflow Cycle Count, Nasgro Long Block Output Format in Stress Time History Input, and Single Axis. Click on Analyze.

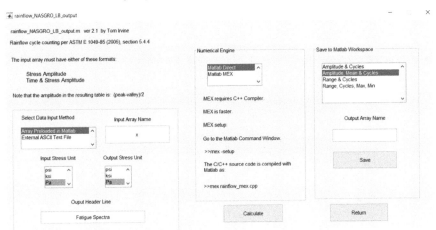

Step 9: Select the data input method (Array Preloaded in Matlab). Enter the array name (Ex: array x is having the time history of the stress). Select the Input and Output stress units as Pascal, and select the Amplitude, Mean, and cycles (to store them in the workspace).

Step 10: This will perform the task of rainflow counting, and the cycle range and the full and half cycles can be seen in range cycles, which is stored in the workspace.

Step 11: To visualize the histogram of stress cycle range and the average cycle stress. The below code is useful.

```
t = seconds(0:0.1:4000)';
%%
TT = timetable(t,x);
rainflow(TT)
hold on
```

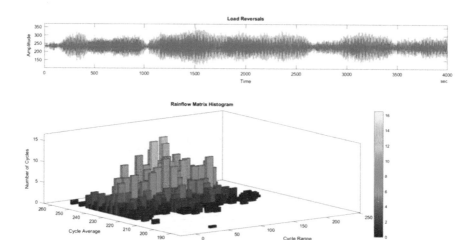

Step 12: Now, each cycle is converted into a non-zero mean cycle using the Goodman diagram. Then, using the S-N curve approach, the number of cycles required to failure is computed. Subsequently, using Miner's rule, damage that occurred to the tethers is computed for the simulation time, and then the fatigue life is determined.

Following is the computer code for computing the damage and fatigue life of mooring lines:

```
% reversible Stress

digitsOld = digits(16);
a=zeros(length(range_cycles),2);
for i=1:length(range_cycles)
    a(i,1)=range_cycles(i,1)/(1-234/670);  %  give  mean
    stress and ultimate tensile stress
    a(i,2)=range_cycles(i,2);
end
%%
% no of cycles req to fail
b=zeros(length(range_cycles),1);

for i=1:length(range_cycles) % adopt the appropriate sn
curve from the code
    if (a(i,1)>=103.8006)
        b(i)=10^(12.049-3*log10(a(i,1)));
    else
        b(i)=10^(16.081-5*log10(a(i,1)));
    end
end
```

```
for i=1:length(range_cycles)
   if(a(i)>41.54)
     b(i,1)=b(i,1);
   else
   end
   end
%% Damage
d=zeros(length(range_cycles),1);
for i=1:length(range_cycles)
   if (a(i,1)>=41.45536426)
     d(i,1)=a(i,2)/b(i,1);
   else
     d(i,1)=0;
   end
   end
D=sum(d);
%% Fatigue Life
L=(1/D)*4000/(365*24*60*60) % change the simulation time
```

Step 13: After obtaining the time history response of tension variation, transfer the data from Excel to MATLAB using the command (readable ('file_name.extension')).

The time history response will get stored in the Matlab Workspace. Then convert the table into an array using the command table2array(table_name).******
Now everything is in an array, and we can proceed for our computations.

1. Now compute the mean and standard deviation using the commands **mean (A(:,3))** and **std(A(:,3))** and store them in **mean_stress** and **std_stress**, respectively.
2. The mean strength of the steel is considered as μ_R = **630.8 N/mm²**.
3. The standard deviation of the steel strength is σ_R = **33.7 N/mm²**.

4. Now, the Reliability Index $= \dfrac{\mu_R - mean\_stress}{\sqrt{\sigma_R^2 + std\_stress^2}} = \dfrac{630.8 - 233.726}{\sqrt{33.7^2 + 45.8629^2}} = \dfrac{397.074}{56.91305} = 6.978.$

5. To compute the probability of failure, we will use the command **normcdf(-β)** = normcdf(-6.978).

Following is the code for the reliability analysis:

```
clear all
clc
A = readtable('Book1.xlsx');
A = table2array(A);
mean_strength = 630.8; %?R mean strength N/mm2
```

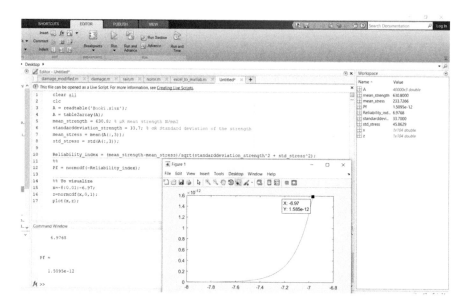

FIGURE 3.79   MATLAB code for reliability analysis.

```
standarddeviation_strength = 33.7; %?R Standard deviation of the
strength
mean_stress = mean(A(:,3));
std_stress = std(A(:,3));
Reliability_index = (mean_strength-mean_stress)/sqrt(standard-
deviation_strength^2 + std_stress^2);
%%
Pf = normcdf(-Reliability_index);
%% To visualize
x=-8:0.01:-6.97;
z=normcdf(x,0,1);
plot(x,z);
```

3.10 ADDITIONAL MATLAB CODES

3.10.1 MATLAB CODE FOR CONVERTING THE NON-ZERO MEAN STRESS TO ZERO MEAN STRESS AND COMPUTING FATIGUE DAMAGE

```
% reversible Stress
a=zeros(length(range_cycles),2);
for i=1:length(range_cycles)
   a(i,1)=range_cycles(i,1)/(1-134/678.1); % give mean stress and
   ultimate tensile stress
   a(i,2)=range_cycles(i,2);
```

```
end
%%
% no of cycles req to fail
b=zeros(length(range_cycles),1);

for i=1:length(range_cycles) % adopt the appropriate sn curve from
the code
   if (a(i,1)>=103.8006)
   b(i)=10^(12.049-3*log10(a(i,1)));
      else
   b(i)=10^(16.081-5*log10(a(i,1)));
end
end
for i=1:length(range_cycles)
   if(a(i)>41.54)
   b(i,1)=b(i,1);
      else
   b(i,1)=0;
end
end
%% Damage
d=zeros(length(range_cycles),1);

for i=1:length(range_cycles)
   if (a(i,1)>=41.45536426)
   d(i,1)=a(i,2)/b(i,1);
   else
         d(i,1)=0;
   end
end
D=sum(d);
%% Fatigue Life
L=(1/D)*4000/(365*24*60*60) % change the simulation time
```

3.10.2 Code for Generating the Rainflow Cycle Counting Graph

```
fs = 39.998;
t = seconds(0:1/fs:1000-1/fs+0.1)';
%%
TT = timetable(t,x);
rainflow(TT)
hold on
```

3.10.3 Code for Obtaining Probability of Failure for a Given Reliability Index

```
x=-10.3:0.01:-10;
%y=normpdf(x,0,1);
z=normcdf(x,0,1);
%plot(x,y);
%hold on
plot(x,z);
```

3.10.4 Code for First-Order Reliability Method

```
clear all  %% iterative method for solving
reliability index for normal distribution
clc
u1=50;
u2=95;
u3=4000;
S1=2.5;
S2=10;
S3=1000;
n=10;
b=zeros(n,1);
b(1,1)=6;
A1=-.58;
A2=-0.58;
A3=0.58;
for i=1:10
  Nr=-2.5*u1*u2+u3;
  Dr1=2.5*A1*A2*S2*S1;
  Dr0=-2.5*A1*S1*u2-2.5*A2*u1*S2+A3*S3;
  Dr=(Dr1*b(i,1)+Dr0);
  b(i+1,1)=Nr/Dr;
  Z1=1.25*(-S1*A2*b(i+1)*S2+S1*u2);
  Z2=1.25*(-A1*b(i,1)*S1*S2+u1*S2);
  Z3=-S3/2;

  K=sqrt(Z1^2+Z2^2+Z3^2);
A1=Z1/K;
A2=Z2/K;
A3=Z3/K;

end
```

Editor - fos.m

11x1 double

	1	2
1	6	
2	-4.4353	
3	4.5154	
4	4.7919	
5	4.7942	
6	4.7941	
7	4.7941	
8	4.7941	
9	4.7941	
10	4.7941	
11	4.7941	
12		
13		

3.10.5 Code for Comparing the Probability of Failure of Correlated and Uncorrelated Random Variables in the Same Distribution

```
clear all
clc
n=10000;
z1 = randn(n,1);
z2 = randn(n,1);
% correlating the random variables
C = [1,0.8;0.8,1];
L=chol(C,'lower');
Z1=z1;
Z2=z1*L(2,1)+z2*L(2,2);
% correlated to given mean and standard
deviation
%M=[12;0.001]
%S=[5,0;0,1000]
%X=[Z1 Z2]*S
%mean(X(:,1))
M1=40;
M2 =30;
S1 =5;
S2 =6;
X1 = Z1*S1+M1;
X2 = Z2*S2+M2;
c=corr(X1,X2)
G=0
% let limit state equation is R-S =0
for i=1:n
   K(i,1)=X1(i,1)-X2(i,1);
   if K(i,1)<=0
     G=G+1;
   end
end
Pf1= G/n
%% For uncorrelated Random Variables
D=0;
R=normrnd(40,5,[n,1]);
S=normrnd(30,6,[n,1]);
for i=1:n
  K(i,1)=R(i,1)-S(i,1);
  if K(i,1)<=0
    D=D+1;
  end
end
pf2=D/n
```

Editor - F:\Matlab_practice\Reliability\probability_failure_correlatedVsUncorrelated

randomnumberstriangulardistribution.m | Linearcongruentialmethod.m

This file can be opened as a Live Script. For more information, see Creating Live Sc

```
1 -    clear all
2 -    clc
3 -    n=10000;
4 -    z1 = randn(n,1);
5 -    z2 = randn(n,1);
6      % correlating the random variables
7 -    C = [1,-0.8;-0.8,1];
8 -    L=chol(C,'lower');
9 -    Z1=z1;
10 -   Z2=z1*L(2,1)+z2*L(2,2);
11     % correlated to given mean and standard deviation
12     %M=[12;0.001]
13     %S=[5,0;0,1000]
14     %X=[Z1 Z2]*S
```

Command Window

```
    0

Pf1 =

    0.1669

pf2 =

    0.1038
```

randomnumberstriangulardistribution.m | Linearcongruentialmethod.m

This file can be opened as a Live Script. For more information, see Creating Live Sc

```
1 -    clear all
2 -    clc
3 -    n=10000;
4 -    z1 = randn(n,1);
5 -    z2 = randn(n,1);
6      % correlating the random variables
7 -    C = [1,0.8;0.8,1];
8 -    L=chol(C,'lower');
9 -    Z1=z1;
10 -   Z2=z1*L(2,1)+z2*L(2,2);
11     % correlated to given mean and standard deviation
12     %M=[12;0.001]
13     %S=[5,0;0,1000]
14     %X=[Z1 Z2]*S
```

Command Window

```
    0

Pf1 =

    0.0024

pf2 =

    0.1017
```

4 Semi-Submersibles under Impact Loads

4.1 INTRODUCTION

As semi-submersibles are one of the commonly preferred compliant platforms for ultra-deep water drilling, the study discussed in this chapter would be useful to understand the structural behavior under accidental impact loads arising due to ship platform collisions. During oil and gas production, a semi-submersible is required to service bigger supply boats and vessels. So, the column members of the platform are prone to impact loads that may arise due to ship-platform collision involving high-impact energy. As a result of such collision events, local or global deformations may occur in the column members, which may subsequently affect the total strength and stability of the whole platform. Also, the response of a semi-submersible under such an accidental load highly depends upon the material and geometric properties of the platform. With the increase in the number of oil production and exploration platforms in deep waters, the risk of ship-collision and ice-impact has also substantially grown in the recent past. The ship-platform collision is a dynamic process, involving several factors, namely, type of collision, contact time, energy absorption, and dissipation, to assess the structural response (Kvitrud, 2011). The compliance of semi-submersibles may even increase the risk involved in impacts, because of very little or even no redundancy in the structure. Also, the post-collapse strength of the main structural components of such platforms would be very low. A relatively small dent of very less thickness would be sufficient enough to violate the safety factor considered in the design of the structural members (Harding et al., 1983).

4.2 MATERIAL PROPERTIES

The structural configuration of a semi-submersible, loading conditions, and material properties make the members different from other commonly used cylindrical shell structures. The impact behavior is strongly dependent on the material properties. In the present discussion, marine steel AH36 is used, whose mechanical properties are listed in Table 4.1 (Cho et al., 2015).

For the numerical study, the true stress-strain values are computed from the engineering stress-strain, as given below (Cho et al., 2015):

$$\sigma_t = \sigma_{eng}(1 + \varepsilon_{eng}) \tag{4.1}$$

$$\varepsilon_t = \ln(1 + \varepsilon_{eng}) \tag{4.2}$$

135

TABLE 4.1

Mechanical properties of marine steel AH36

Mechanical properties	Value	Units
Yield strength	433	N/mm^2
Young's modulus	2.06 x 10^5	N/mm^2
Ultimate tensile strength	547	N/mm^2
Ultimate tensile strain	0.156	No unit
Hardening start strain	0.0214	No unit
Ultimate tensile strain	0.1692	No unit

where σ_t is the True stress, σ_{eng} is the engineering stress, ε_t is the true strain, and ε_{eng} is the engineering strain. The static constitutive equation considering the yield plateau is given by:

$$\sigma_t = E\varepsilon_t, 0 < \varepsilon_t \leq \varepsilon_{Yt} \tag{4.3}$$

$$\sigma_t = \sigma_{Yt} + (\sigma_{HSt} - \sigma_{Yt})\left(\frac{\varepsilon_t - \varepsilon_{Yt}}{\varepsilon_{HSt} - \varepsilon_{Yt}}\right), \varepsilon_{Yt} < \varepsilon_t \leq \varepsilon_{HSt} \tag{4.4}$$

$$\sigma_t = \sigma_{HSt} + \left(\frac{\sigma_{Tt} - \sigma_{HSt}}{(\varepsilon_{Tt} - \varepsilon_{HSt})^A}\right)(\varepsilon_t - \varepsilon_{HSt})^A, \varepsilon_{HSt} < \varepsilon_t \tag{4.5}$$

$$A = \frac{\sigma_{Tt}}{\sigma_{Tt} - \sigma_{HSt}}(\varepsilon_{Tt} - \varepsilon_{HSt}) \tag{4.6}$$

where σ_{Yt} is the true yield stress, σ_{HSt} is the true hardening start to stress, σ_{Tt} is the true ultimate stress, ε_{Yt} is the true yield strain, ε_{HSt} is the true hardening start to strain, and ε_{Tt} is the true ultimate strain. The true stress-strain curve for AH36 steel, considering the yield plateau, is shown in Figure 4.1, which represents the state of the material more accurately. Please note that the initial yield delay is neglected while developing the true stress-strain curve in the constitutive equations. To better predict the impact response, the effect of the strain rate is considered in the analysis. It provides an improved assessment as it includes the residual strength under damaged conditions against the impact load (Cerik et al., 2015). When the strain rate increases, yield delay occurs in the material before entering the yield plateau; the width of the yield plateau is also affected by the strain rate.

Dynamic constitutive equations include the effects of strain-rate hardening and give an improved assessment of structural response:

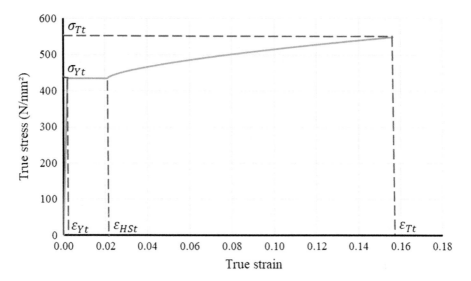

FIGURE 4.1 True stress-strain curve of AH 36 grade steel.

$$\sigma_{Yd} = 1 + \left\{ \left(\frac{E\sigma_T}{1000\sigma_Y^2} \right)^{4.89} \left(\frac{\dot{\varepsilon}}{233.6} \right) \right\}^{0.333} \tag{4.7}$$

$$\sigma_{Td} = (1 + 0.12\dot{\varepsilon}^{(-n)})\sigma_{Yd} \tag{4.8}$$

$$\varepsilon_{HSd} = (1 + (p \times \dot{\varepsilon})^{0.3})\varepsilon_{HS} \tag{4.9}$$

$$\varepsilon_{Td} = \left(1 + \frac{\dot{\varepsilon}}{q} \right)^{-0.333} \tag{4.10}$$

$$n = 11.14 \times \exp\left(\frac{138\sigma_Y}{\sigma_T} \right) \tag{4.11}$$

$$p = 10.2 \times \exp\left(-0.52 \left(\frac{1000\sigma_Y}{E} \right)^{2.2} \left(\frac{\sigma_T}{\sigma_Y} \right)^{2.8} \right) \tag{4.12}$$

$$q = 4.71 \left\{ \left(\frac{1000\sigma_Y}{E} \right)^{8} \left(\frac{\sigma_T}{\sigma_Y} \right)^{0.3} \right\}^{2.08} \tag{4.13}$$

where $\dot{\varepsilon}$ is the strain rate, σ_{Yd} is the dynamic yield strength, σ_{Td} is the ultimate dynamic strength, ε_{HSd} is the dynamic hardening start to strain, and ε_{Td} is the dynamic ultimate tensile strain. From the dynamic constitutive equation, the true stress-plastic strain curves are plotted for different strain-rate values ($10\ \mathrm{s}^{-1}$, $25\ \mathrm{s}^{-1}$, $50\ \mathrm{s}^{-1}$, $75\ \mathrm{s}^{-1}$, $100\ \mathrm{s}^{-1}$, $125\ \mathrm{s}^{-1}$, $150\ \mathrm{s}^{-1}$), as shown in Figure 4.2. It is seen that the initial yield and the width of the yield plateau increase with the increase in the strain rate. At higher strain rates, the effect of strain hardening is reduced significantly. Also, the material behaves perfectly plastic under an increased strain rate. Note that the above equations provide better results than the Cowper-Symonds equation (Singh et al., 2011).

4.3 SHIP-PLATFORM COLLISION

Standard regulations and guidelines for the design of offshore structures help in reducing the failure of platforms under accidental loads. Based on NORSOK N-003 guidelines for production platforms, 5000-ton supply ships with speeds not lesser than 2.0 m/s should be considered for the design check under impact. It is important to note that design codes permit a level of moderate to significant damage to the platform, but with a limitation; it should not cause structural disintegration and lead to progressive collapse of the structure. Further, the Norwegian Maritime Directorate regulation suggests a design check under the collision of a 5000-ton ship at a speed of about 2.0 m/s. The design guidelines suggested a minimum of 4-MJ collision energy for the design of offshore structures under accidental events (Table 4.2).

Based on a statistical overview, it is observed that the maximum number of collisions occurred with the visiting vessels. These vessels may be either patrolling

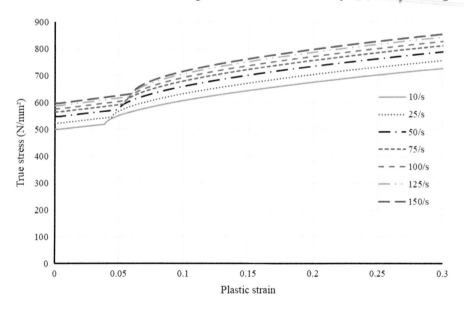

FIGURE 4.2 True stress-plastic strain plot for different strain rates.

TABLE 4.2

Ship-platform collision

Platform	Vessel	Date	Mass of vessel	Speed of vessel (m/s)	Collision energy (MJ)
West Venture semi-submersible	Far Symphony	7 March 2004	5000 tons	3.7	39
Ekofisk 2/4-P Jacket	Ocean carrier	2 June 2005	4679 tons	3	20
Njord B FSU	Navion Hispania	13 November 2006	126.13 tons	1.2	61
Grace Jacket	Bourbon Surf	18 June 2007	3.117 tons	1–3.5	Insignificant
Ekofisk 2/4-P tripod Jacket	Big Orange XVIII	8 June 2009	6000 tons	4.5–4.8	70

vessels, oil tankers, transport vessels, or even vessels anchored to the platform for emergency rescue. An average size of such vessels is about 100 tons; in the past, the average size was 20 tons. This increase is due to the large size of operating vessels, which are in increasing demand during remote operations in the deep-sea. With such an increased weight and size of the abutting vessels, it is obvious to note that offshore platforms are subjected to more possible damage due to the increase in the collision energy. Major collision events that occurred in the recent past are listed in Table 4.1. It is evident from the table that the design collision events considered in standard regulations are less probable; recent collision events occurred with greater collision energy and increased speed of the vessel. The collision cases considered in this study are given in Table 4.3 (Syngellakis and Balaji, 1989).

To investigate the response of the column members of a semi-submersible under impact loads, a rectangular, box-shaped indenter of size (10 x 5 x 2 m) with a displacement of 7500 ton is considered as a striking mass. It represents the ship stem, which is also called the stem bar. Please note that the displacement of the indenter considered in this study is 50% more than the design guidelines. It is not

TABLE 4.3

Collision speed and impact duration

Case	Collision speed (m/s)	Impact duration (seconds)
Case 1	1.0	0.30
Case 2	2.0	0.35
Case 3	3.0	0.38
Case 4	4.0	0.40

(a) Vessel impact with semi-submersible

(b) Cylindrical shell and location of indenter

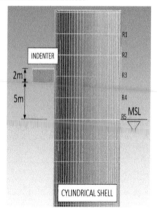

FIGURE 4.3   Schematic view of column member and indenter.

only to envisage a conservative approach but also due to the increase in size of the offshore support-vessels in the recent past. Figure 4.3 shows a schematic view of the indenter encountering the column member.

Ring stiffeners, provided at various locations along the length of the member, are indicated as R1, R2, R3, R4, and R5 in Figure 4.3; vertical lines indicate stringers. The indenter impacts the column at a height of 5 m above the Mean Sea Level (MSL). The assumed load cases, as shown in Table 4.3, are considered to be a central, sideways collision of the ship with the column of a semi-submersible. The impact load caused by the indenter is classified as a low-velocity impact because the velocity of impact is much less than the conventional impact velocity of 20 m/s. In the analysis, it is assumed that the indenter is infinitely rigid, and the energy is dissipated only by the semi-submersible platform. A ductility-based design is followed, which implies that the platform dissipates the major part of the collision energy by undergoing large plastic deformation (Det Norske Veritas, 2016a,b; 2008).

4.4 RESPONSE UNDER IMPACT LOADS

The column member of the semi-submersible platform is modeled as an ortho-gonally stiffened cylindrical shell of diameter equivalent to that of the square section. The central-difference, time-integration scheme is used as the solver. A rectangular box-shaped indenter, resembling the stem of the ship, is modeled using solid elements and assumed to be perfectly rigid. Thus, the strain energy dissipation is confined to the cylindrical shell and the stiffeners during impact (Silvestere, 2008; Tennyson, 1971). The contact region is defined using the general body in-teractions option in the solver. The outer surface of the shell and the indenter surface are chosen as the contact surfaces. The Courant-Friedrichs-Lewy condition is followed to limit the time step used in the explicit analysis. It is also useful to ensure the stability and accuracy of the solution. The column member is modeled as a four-node, quadrilateral shell element. Nodes on the sides of the shell elements are connected by the vertical lines. A regular pattern is used to mesh the indenter. The quality of the solution is checked through momentum and energy conservation for different mesh sizes. A mesh size of 0.3 m is found to be adequate in predicting the stress-strain relationship accurately. Discontinuities that arise in the flow variables under shock waves that are caused by the strong impact are handled by viscous terms in the solver. A quadratic, artificial viscosity coefficient of unity is used to avoid discontinuities. Further, a linear, artificial viscosity coefficient of 0.2 is used to damp the oscillations in the solution. The striking mass is restrained in all de-grees-of-freedom except in the impact direction. The initial collision velocity of 5 m/s is applied to the rectangular indenter in the impact direction for simulating the desired impact energy.

The material plasticity data, as shown in Figures 4.1 and 4.2 are used to define the true stress-equivalent plastic strain, obtained from the constitutive equations. A multi-linear, isotropic-hardening model with Von-Mises failure criteria is used in the numerical simulation. As shown in Figure 4.3, the horizontal and vertical lines observed on the surface of the cylindrical shell are stringers and ring-stiffeners, respectively. With due consideration to the height of the visiting vessels in the offshore industry, the indenter is placed at about 5.0 m above the MSL. Previous studies reported that the predominant component of deformation is the local in-dentation at the initial stages of the impact loading. To compensate for an increase in the bending rigidity, the ends of the cylindrical shell are modeled as simply-supported ends. The above idealization reduces both the run-time and the compu-tational effort during the analysis. However, fine optimum meshing adopted in the study had considerably extended the runtime.

4.4.1 DAMAGE PROFILE

The indenter hits the column at 5.0 m above the MSL, where the ring stiffener R3 is located. The indenter velocity and displacement, along the direction of impact under different impact load cases, are shown in Figures 4.4 and 4.5. The equivalent stress distribution and the deformed shape of the column under impact load case-1 and case-2 are shown in Figures 4.6 and 4.7, respectively.

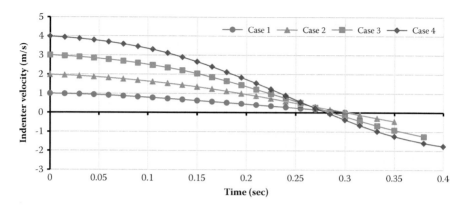

FIGURE 4.4 Indenter velocity for different impact cases.

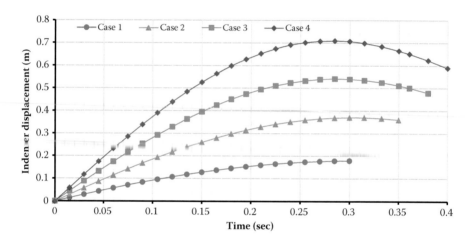

FIGURE 4.5 Indenter displacement for different impact cases.

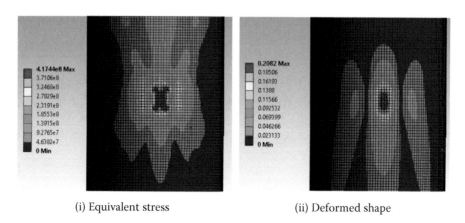

(i) Equivalent stress   (ii) Deformed shape

FIGURE 4.6 Damage profile under impact load case-1.

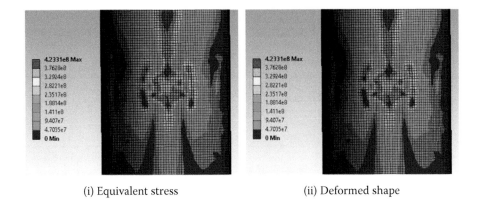

(i) Equivalent stress   (ii) Deformed shape

FIGURE 4.7   Damage profile under impact load case-2.

As observed from the figures above, the impact of the indenter causes a local dent, leading to flattening of the outer cylindrical shell of the column member and the ring stiffener at the impact location. It is also seen that flattening of the local dent increases with the increase in the contact area. Ring stiffeners restrain the spread of damage to the adjacent bays as they obstruct the circumferential bending (Do et al., 2018). The ring stiffener (R3) undergoes the maximum deformation, whereas the adjacent ring frames deformed only about 35% of that of R3. Ring frames, provided at the ends of the cylindrical shell, remain circular, and they are unaffected by the impact load. The maximum strain in the cylindrical shell is observed within the proximity of the ring stiffener (R3). The stringer provided near the impact location collapsed as a beam, spanning between the ring stiffeners. With the increase in the dent depth at the impact location, adjacent stringers also deform jointly with the cylindrical shell. Further, local tripping of stringer stiffeners is also observed closer to the deformed ring stiffener. With the increase in load case-2 (impact velocity is 2 m/s), an increase in the equivalent stress is observed in the cylindrical shell from its top end to the impact location. However, the ring frames reduced the spread of deformation to the adjacent bays from the impact location. The maximum strain is observed only on the ring stiffener at the impact location. An increase in strain is observed up to a depth of about 6.0 m below the MSL. Twisting in the stringers is also observed very close to the ring stiffener (R3) in the impact location.

Figures 4.8 and 4.9 show the equivalent stress and deformed shape under impact cases 3 and 4, respectively. It is observed that with the increase in the impact velocity and duration, the maximum equivalent stress and deformation increases in the cylindrical shell. The principal stress in the cylindrical shell increases up to a depth of about 27 m below the MSL. Under case-4, the damage causes the cylindrical shell to bulge out excessively at the end of the flattened dent. Plastic strain is seen only at the impact location without spreading in the longitudinal direction. It can be also noticed that the plastic strain is greater than that of the elastic strain. Under case-4, the equivalent stress in the member increases beyond its yield stress capacity, which resulted in the reduction of the load carrying capacity of the column member.

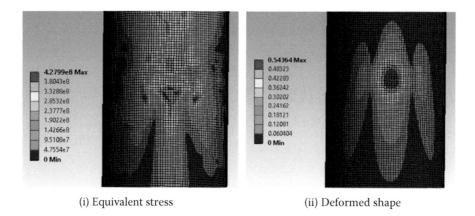

(i) Equivalent stress   (ii) Deformed shape

FIGURE 4.8   Damage profile under impact load case-3.

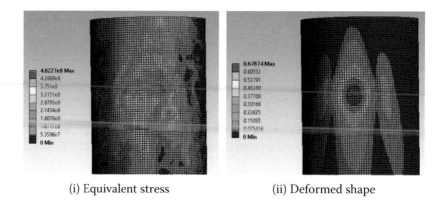

(i) Equivalent stress   (ii) Deformed shape

FIGURE 4.9   Damage profile under impact load case-4.

The deformation pattern of ring stiffener R3, as seen in Figure 4.10, is quite interesting. With the increase in the contact area, flattening of the cylindrical shell increases. It also bulges out from the ends of a flattened section along with the ring stiffener, which is present closer to the impact location. Because ring stiffener R3 is located near the collision zone, the maximum plastic strain is observed only on this stiffener under all impact load cases.

4.4.2 FORCE-DEFORMATION

The results of the impact analysis are summarized in Table 4.4. As seen from the table, the peak force and deformation in the column member increase with the increase in the impact velocity and duration. The force versus non-dimensional deformation curves are shown in Figure 4.11. The area under these curves shows the total energy absorbed by the column under different impact load cases. As seen from the figure, the flattening of curves at a particular instant of time occurs due to

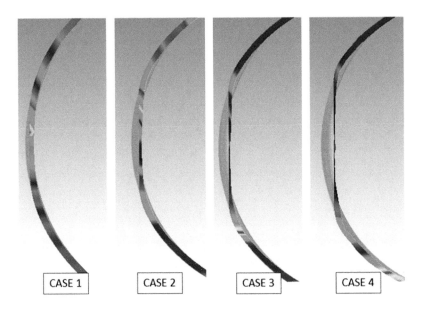

FIGURE 4.10   Deformation of ring stiffener R3.

TABLE 4.4
Summary of impact analysis

Case	Peak force (MN)	Shell deformation (m)	Maximum energy absorbed (MJ)
Case 1	2.189	0.208	0.392
Case 2	4.123	0.420	1.568
Case 3	6.519	0.587	3.532
Case 4	8.823	0.758	6.279

the torsional buckling of the stiffeners. It can be stated that the impact loads with higher intensity lead to the local weakening of the member. The energy absorbed by the column member under different impact loads is shown in Figure 4.12. It is seen from the plots that the maximum energy absorbed under very high impact velocity is about 6.3 MJ.

4.5 PARAMETRIC STUDIES ON IMPACT RESPONSE

4.5.1 LOCATION OF COLLISION ZONE

The vessel draft, wave height, and maximum tide level decide the collision zone on the column member. Even a small change in the impact location may affect the response significantly. It is due to the fact the behavior is morelocalized within the bay between two ring stiffeners, but may also gets transferred to the adjacent

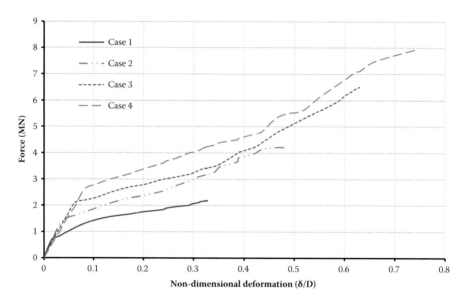

FIGURE 4.11 Load versus non-dimensional deformation curve of column member.

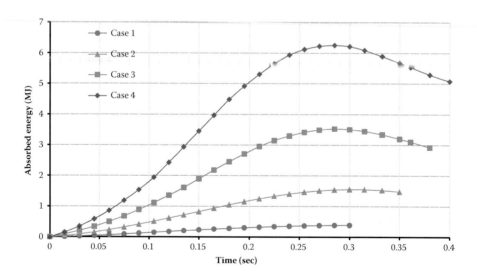

FIGURE 4.12 Energy absorbed by column member under impact load.

stiffener, as seen in the above section. Considering a single impact velocity of 4 m/s, the column member is analyzed to estimate the extent of the collision zone. The location of the rectangular indenter varies along the length of the column member above the MSL as shown in Figure 4.13. The depth of the indenter is 2.0 m, and the location of the center of the indenter is represented as the impact locations. Ring stiffeners R3 and R4 are located at the impact locations 1 and 5, respectively. The impact load under location-3 acts on the mid-bay between ring stiffeners R3 and R4.

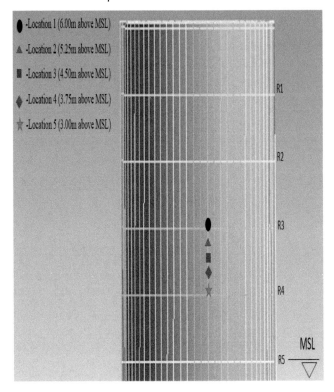

FIGURE 4.13 Location of indenter.

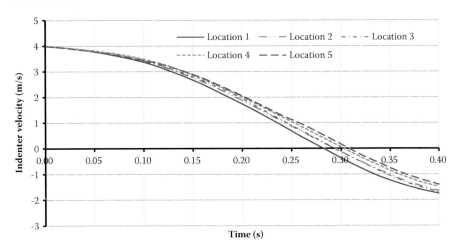

FIGURE 4.14 Indenter velocity for various impact locations.

Variations in the indenter velocity and displacement for different impact locations are shown in Figures 4.14 and 4.15, respectively. It is observed that a reduction in the indenter velocity during impact is not uniform at different impact locations. It is due to the variation in the resistance offered by the cylindrical shell and stiffeners.

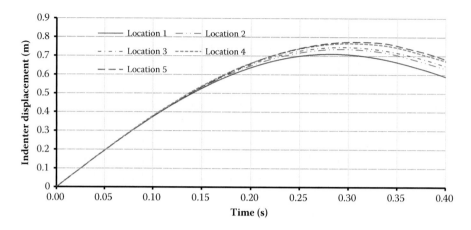

FIGURE 4.15   Indenter displacement for various impact locations.

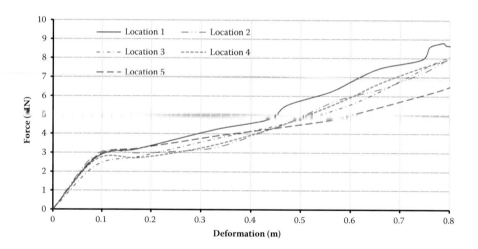

FIGURE 4.16   Force-deformation curves at different impact locations.

A marginal variation in the indenter displacement is also observed. However, the displacement of the indenter during the initial stages of impact is the same at different locations. It is due to the uniform resistance offered by the cylindrical shell at the initial stages of impact.

The force-deformation curves of the column member at different impact locations are shown in Figure 4.16. It is seen from the plot that with the increase in the distance of the impact location, measured from the top end of the member, maximum deformation occurs under lesser impact force. At impact locations 1 and 5, the ring stiffener undergoes maximum deformation compared to that of the stringers and cylindrical shell. Though the center of the indenter does not coincide with the

TABLE 4.5
Impact test results for different impact locations

Location	Peak force (MN)	Maximum shell deformation (m)	Maximum ring frame deformation (m)	Maximum stringer deformation (m)	Maximum energy absorbed (MJ)
1	8.823	0.759	0.845	0.788	6.279
2	8.525	0.784	0.837	0.813	6.277
3	8.390	0.784	0.769	0.844	6.274
4	8.186	0.803	0.796	0.829	6.276
5	8.086	0.805	0.834	0.957	6.271

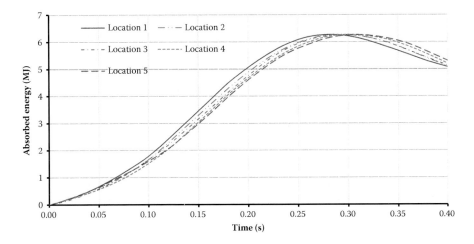

FIGURE 4.17   Energy absorbed from impact force under different locations.

location of the ring stiffener at impact locations 2 and 4, the global deformation in the ring stiffener is comparatively higher than that of the cylindrical shell and stringers. Under impact locations 2 and 4, yielding occurs initially in the stringers and then gets transferred to the ring frames. However, the ring frames play a significant role in resisting the impact force under all cases.

Under the mid-bay impact, stringers yield like a beam, which is restrained at its ends by the ring stiffeners. Under this condition, stringers yield first and then transfer the load to the ring frames. This leads to an increase in the global deformation of the shell. Despite the distinct deformation pattern at impact locations 1 and 3, the variation in the impact force on the column member is only marginal. The summary of the test results is given in Table 4.5. It can be seen that, under all cases, the ring stiffener that is closer to the impact location undergoes the maximum deformation. The maximum strain is observed only in ring stiffeners R3 and R4 at impact locations 1 and 5, respectively. Stringers that are closer to the deformed ring stiffeners are also subjected to the maximum strain. This validates the anology that the response of the column member is highly affected by the change in impact location. The energy absorbed by the buoyant leg due to the impact force under different locations is shown in Figure 4.17.

Although variations in the energy absorbed by the column member under different impact locations is marginal, significant changes in the force-deformation curves are observed due to the differential yielding of the cylindrical shell and the stiffeners. With the increase in the distance of positioning of the indenter from the top end of the cylindrical shell, the impact duration at which the member absorbs the maximum energy also increases. But the magnitude of maximum absorbed energy remains the same under different impact load cases. Out of all the impact locations assessed, location-5 seems to be critical as it causes the maximum deformation under a lesser impact force (see Figure 4.13).

The deformed shape of the cylindrical shell and ring stiffener R3 for different impact locations of the indenter are shown in Figures 4.18 and 4.19, respectively. As seen from the plots, the maximum deformation of the stiffener significantly changes with the indenter position; it is reduced with an increase in distance. It is also important to observe that the ring stiffener does not limit the damage of the shell in the mid-bay. Under mid-bay collision, ring stiffeners, located at the ends, also undergo tilting with the change in the circumferential strain. It increases when the longitudinal damage from the mid-bay reaches the ring stiffener. Ring frames also deform laterally, and the flattened zone starts with the beam action, supported by the shell at both ends. Increased deformation in the cylindrical shell provokes compression and finally leads to tilting of the ring stiffeners. Then, the damage starts propagating further in the longitudinal direction.

Irrespective of the location of the indenter, ring stiffeners play a major role in energy dissipation. In the mid-bay collision, stringer stiffeners resist the impact load by beam action; beam collapse results in the formation of a plastic hinge. Tripping of stringer stiffeners is also seen at a closer proximity to the ring stiffeners. Thus, stringer stiffeners play a significant role in resisting the impact load under impact at the mid-bay. Ring stiffeners obstruct the spreading of damage to the adjacent bays.

4.5.2 Size of Indenter

Numerical analyses are carried out by varying the depth of the rectangular indenter from 1 m to 3 m for the impact velocity of 4.0 m/s as shown in Figure 4.20. With the change in the depth of the indenter, the contact area and the impact force also change. As the impact behavior and deformed profile of the column member showed a significant variation with respect to the location of the indenter, a parametric study is carried out at location 1 (on ring stiffener) and location 3 (mid-bay). The impact load cases and the results are summarized in Table 4.6. The maximum deformation of the column member is seen under collision at the mid-bay for different sizes of the indenter. Impact load cases are represented by the b/R ratio, where b is the depth of the indenter and R is the outer radius of the cylindrical shell.

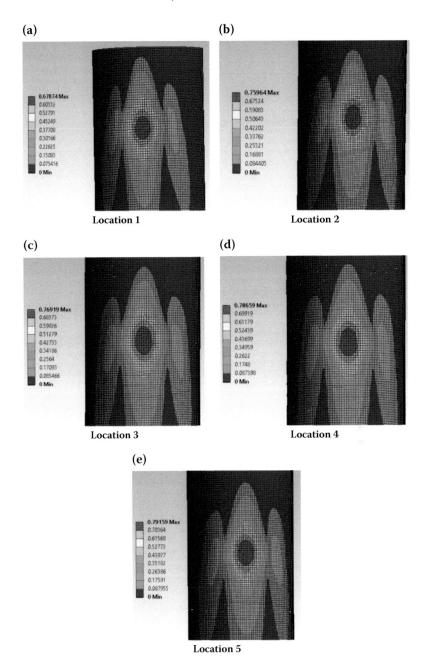

FIGURE 4.18 Deformed shape of cylindrical shell for various indenter locations.

Force-deformation plots under different cases for impact at locations 1 and 3 are shown in Figures 4.21 and 4.22, respectively. It is seen from the plots that the size of the indenter affects the impact response characteristics, significantly. Energy absorbed

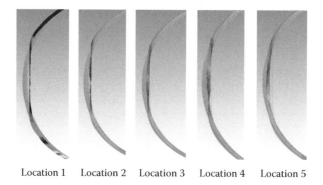

Location 1   Location 2   Location 3   Location 4   Location 5

FIGURE 4.19   Deformation of ring stiffener R3 for various indenter locations.

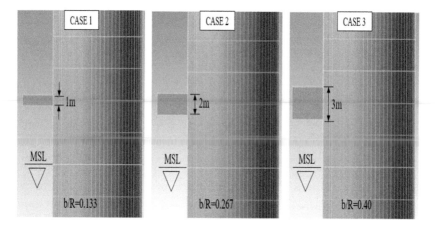

FIGURE 4.20   Impact cases for different size of indenter.

TABLE 4.6

Impact response under different sizes of the indenter

Location	b/R ratio	Peak force (MN)	Maximum deformation (m)	Maximum energy absorbed (MJ)
1 (Ring	0.133	11.696	0.641	6.268
stiffener 3)	0.267	8.82	0.789	6.279
	0.40	7.688	0.923	6.271
3 (mid-bay)	0.133	10.742	0.676	6.28
	0.267	8.39	0.838	6.274
	0.40	7.348	0.957	6.276

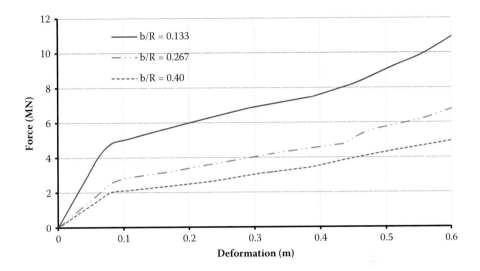

FIGURE 4.21   Force-deformation at location-1 for different sizes of indenter.

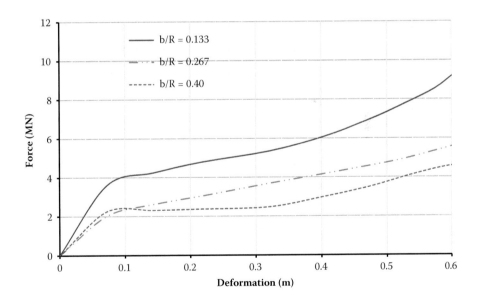

FIGURE 4.22   Force-deformation at location-3 for different sizes of indenter.

by the column member at both the locations are plotted in Figures 4.23 and 4.24, respectively. It is seen from the figures that a higher impact force is developed by the indenter with a lower (b/R) ratio of 0.133. The slope of the energy curve is smaller for indenters with a higher (b/R) ratio in both the locations (1, 3). It confirms that the indenters with greater depth are more critical and capable of causing more deformation even under a smaller impact force.

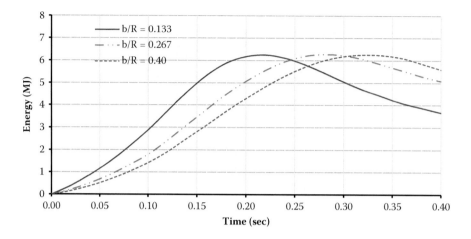

FIGURE 4.23   Energy absorbed from impact force in location-1.

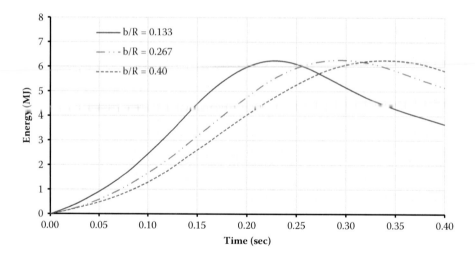

FIGURE 4.24   Energy absorbed from impact force in location-3.

Figures 4.25 and 4.26 show the deformed shape of the column member at locations 1 and 3, respectively. The plots show that with the increase in the (b/R) ratio, there is an increase in the depth of the dent on the column member at both the impact locations. The dent depth increases with the increase in the b/R ratio due to the increased area of contact at both locations. It is also observed that the pattern of damage in the column member is not influenced by the size of the indenter. But the extent of damage along the length of the member and bulging of the member increase as the depth of the indenter increases. Figure 4.27 shows the deformed shape of the ring stiffener at two locations, namely, 1 and 3. It is seen that the ring stiffener is affected more at location-1, causing tripping of the stringers as well.

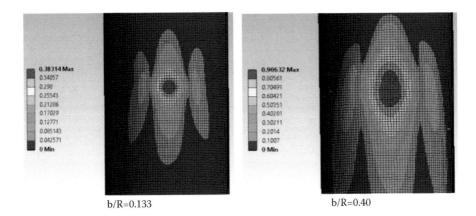

b/R=0.133   b/R=0.40

FIGURE 4.25   Deformed shape of column member at location-1.

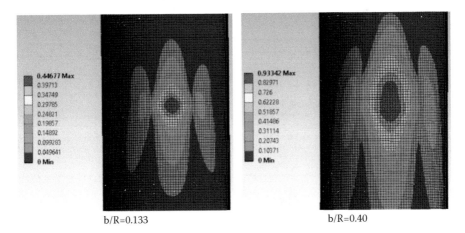

b/R=0.133   b/R=0.40

FIGURE 4.26   Deformed shape of column member at location-3.

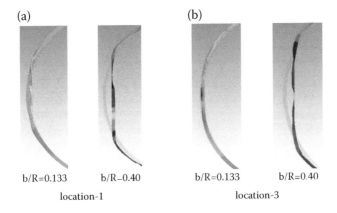

(a)   (b)

b/R=0.133   b/R–0.40   b/R=0.133   b/R=0.40

location-1   location-3

FIGURE 4.27   Deformed shape of ring stiffener.

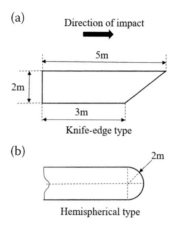

FIGURE 4.28   Shape of indenter.

TABLE 4.7
Impact test results for different indenter shapes

Type of indenter	Knife edge		Hemisphere	
	Location 1	Location 3	Location 1	Location 3
Peak force (MN)	9.5	8.374	10.982	10.61
Max. deformation (m)	0.811	0.774	0.601	0.632
Max. energy absorbed (MJ)	6.263	6.277	6.267	6.267
Ring deformation ratio	0.964	0.766	1	0.571
Stringer deformation ratio	1	1	0.967	1

4.5.3 SHAPE OF INDENTER

A rectangular-shaped indenter is used to represent the stem of the ship (Khedmati and Nazari, 2012). Studies are carried out to understand the influence of the shape of the indenter on the column member under impact locations, namely, 1 and 3. Two different shapes, namely, knife-edge and hemispherical types, are considered, as shown in Figure 4.28. In both cases, the impact velocity of 4.0 m/s is maintained.

The load cases with results are given in Table 4.7.

The deformation ratio, as expressed in the above table, is the ratio of deformation of the respective element and that of the column member. The force-deformation curves of the knife edge indenter and hemispherical indenter are shown in Figures 4.29 and 4.30, respectively. As seen from the pattern of force-deformation curves, flattening of the curve is high under the impact of the knife edge indenter. The knife edge indenter also causes maximum deformation in the stringers. Because of the reduced cross-sectional area, the hemispherical indenter does not cause flattening of the ring stiffener and cylindrical shell. It causes maximum deformation in the ring stiffener and stringer at locations 1 and 3, respectively. It validates the fact that the high impact pressure over small

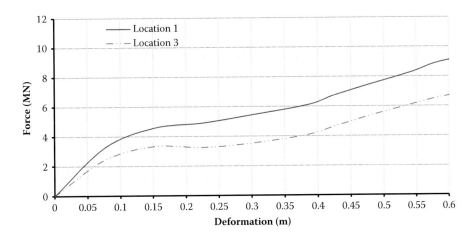

FIGURE 4.29   Force-deformation curves for knife-edge indenter.

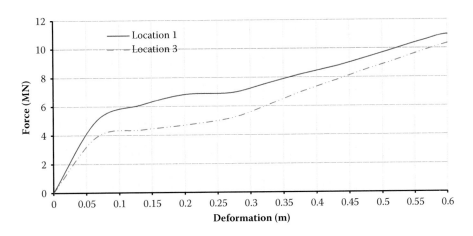

FIGURE 4.30   Force-deformation curves for hemispherical indenter.

areas causes maximum deformation in both stiffeners and the shell. Under all types of indenters, the deformation starts at the flattening of circumferential curvature; it further leads to compressive strain in the membrane. Because of the reduced resistance offered by the stiffeners at the mid-bay, deformation is maximum in both the cases.

The deformed shape of the column member under the impact of the knife-edge and hemispherical indenter are shown in Figures 4.31 and 4.32, respectively. It is seen from the plots that the knife-edge indenter at the mid-bay causes more flattening of the cylindrical shell at the impact location. It also shows higher bulging at the ends of the flattened section. It confirms that the indenters with pointed ends may cause severe local damage to the column members. A reduced area of contact of the hemispherical indenter reduces the flattening of the shell, but distorts it by increasing the circumferential strain. Figure 4.33 shows the deformed shape of the ring stiffener under the impact of two shapes of the indenter. It is seen from the plots that stiffeners

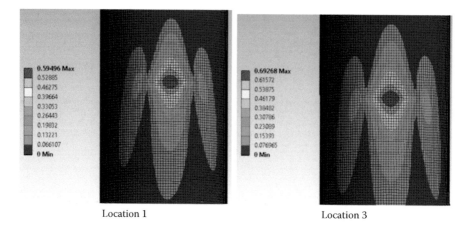

Location 1   Location 3

FIGURE 4.31   Deformed shape of column under knife-edge indenter.

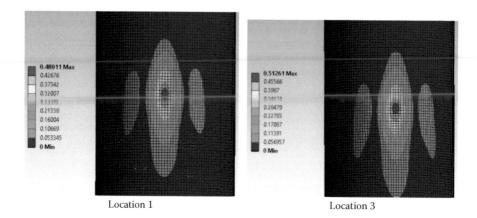

Location 1   Location 3

FIGURE 4.32   Deformed shape of column under hemispherical indenter.

in the impact location are heavily damaged in this case. Comparatively, flattening of the ring stiffener is higher under the impact of the knife-edge indenter. The hemispherical indenter also causes distortion, which can be seen from the distorted profile at the impact location. Spread of damage in the longitudinal direction is comparatively less with the hemispherical indenter, where the damage is highly concentrated only at the impact location. Irrespective of the shape of the type of indenters, stiffeners always play a major role in impact resistance. It, therefore, becomes an important aspect of the design of column members in a semi-submersible.

4.5.4 Strain-Rate Hardening

The Cowper-Symonds equation is commonly used for defining the strain-rate hardening characteristics of the material. However, the coefficients used in the

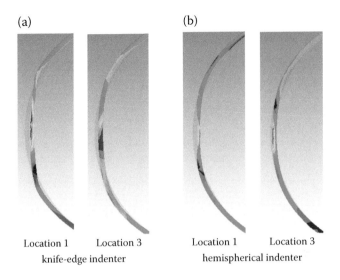

(a)   (b)

Location 1   Location 3   Location 1   Location 3
knife-edge indenter   hemispherical indenter

FIGURE 4.33 Deformation of ring stiffener.

equation vary with strain magnitude. Further, the scaling of quasi-static plasticity data with the dynamic hardening factor will also be insufficient. Hence, the material constant coefficients in the Cowper-Symonds equation are evaluated and then impact analysis is carried out to study the effect of strain-rate hardening on the impact response of the platform. The dynamic hardening factor, which is the ratio of

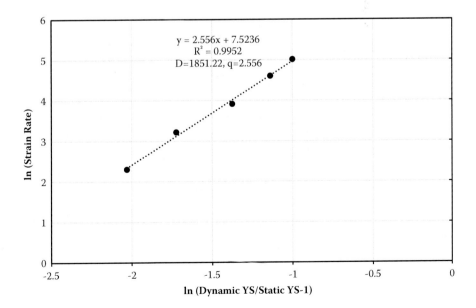

$y = 2.556x + 7.5236$
$R^2 = 0.9952$
$D = 1851.22, q = 2.556$

ln (Strain Rate)

ln (Dynamic YS/Static YS-1)

FIGURE 4.34 Model parameters.

dynamic yield strength to static yield strength scaled along with the quasi-static plasticity data, is used to define the strain-rate effect. It is also assumed that the stress-strain dependence is similar at all strain-rate levels. The Cowper-Symonds equation is given by,

$$\frac{\sigma_D}{\sigma_s} = 1 + \left(\frac{\varepsilon_p}{D}\right)^{\frac{1}{q}} \qquad (4.14)$$

where σ_D is the dynamic yield stress, σ_s is the quasi-static yield stress at 0.001/s strain rate, ε_p is the equivalent strain rate, and D and q are the material constants. Based on the strain-rate hardening detail, the stiffness of the stiffened cylindrical shell will get altered, which in turn affects the plastic strain in the model. The material constants D and q of AH36 marine steel are obtained from the dynamic hardening factor vs strain-rate plot as shown in Figure 4.34.

Based on the derived material constants (D=1851.22, q=2.556), the equation can be written as,

$$\sigma_D = 437\left[1 + \left(\frac{\varepsilon_p}{1851.22}\right)^{\frac{1}{2.556}}\right] \qquad (4.15)$$

The yield surface of the isotropic strain hardening for strain-rate dependent materials is defined as follows:

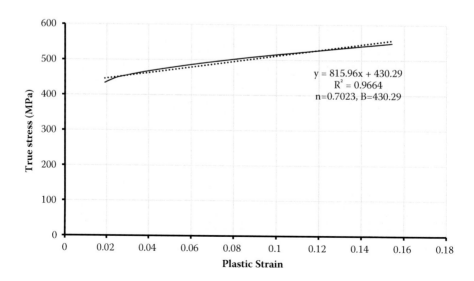

FIGURE 4.35 Determination of model parameters.

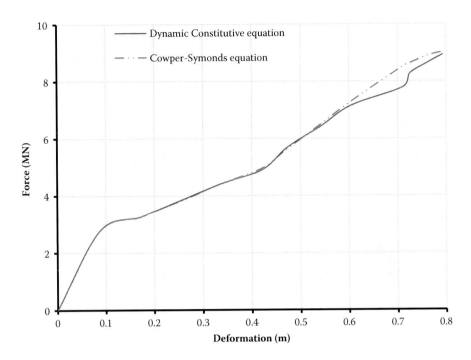

FIGURE 4.36   Force-deformation curves for 25/s strain rate.

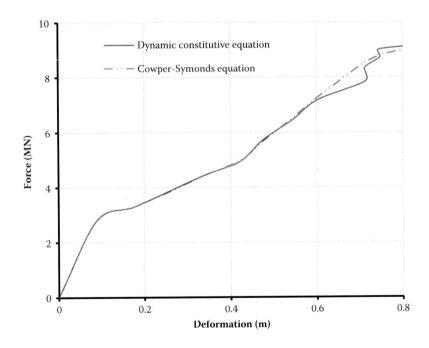

FIGURE 4.37   Force-deformation curves for 50/s strain rate.

$$\sigma_D = (A + B\varepsilon^n)\left[1 + \left(\frac{\varepsilon_p}{D}\right)^{\frac{1}{q}}\right] \qquad (4.16)$$

where A is the yield stress at zero plastic strain, B is the strain hardening coefficient, and n is the strain hardening exponent. The material parameters B and n are calculated by the curve fitting method as shown in Figure 4.35 in the plastic strain versus true stress plot (Singh et al., 2011). Numerical analyses under strain rates of 25/s and 50/s are then carried out using the material definition from Cowper-Symonds model.

Force-deformation curves obtained from the numerical investigations of different strain-rate hardening definitions are shown in Figures 4.36 and 4.37 for strain rates 25/s and 50/s, respectively. The overall response under both the cases does not show significant variations as the constants used in Cowper-Symonds model are suitably modified. It confirms the use of dynamic material properties to define plasticity at higher strain rates. However, marginal variations observed at the higher impact duration decide the maximum permanent deformation on the cylindrical shell and the stiffeners.

Therefore, for a precise prediction of impact response, the effect of strain rate becomes imperative. Numerical studies are carried out by varying the strain-rate effect under different impact load cases. Material properties are included to account for the strain-rate effect in the analysis. It is seen that the effect of strain hardening decreases with an increase in the strain rate. Hence, strain rates of 25/s, 50/s, and 100/s are considered for the parametric study and the results are tabulated in Table 4.8. It is observed that the effect of the strain rate on the impact

TABLE 4.8
Effect of strain-rate hardening

Case	Strain rate (/s)	Peak force (MN)	Shell deformation (m)	Maximum energy absorbed (MJ)
Case 1	25	2.222	0.228	0.392
	50	2.224	0.229	0.392
	100	2.225	0.229	0.392
Case 2	25	4.281	0.412	1.570
	50	4.310	0.412	1.570
	100	4.310	0.415	1.570
Case 3	25	6.689	0.571	3.521
	50	6.712	0.571	3.521
	100	6.757	0.573	3.528
Case 4	25	9.075	0.730	6.263
	50	9.125	0.730	6.268
	100	9.167	0.731	6.273

TABLE 4.9
Mechanical properties of DH36 steel

Temperature	Yield strength (N/mm²)	Ultimate strength (N/mm²)
Room Temperature-RT	383.7	530.2
Arctic Temperature-AT (−60°C)	446.2	606.5

response of the column member is much less. It is also seen that the strain-rate effect increases the peak force acting on the column member at higher impact velocities. However, the increase in peak force is very marginal (about 1%) and therefore insignificant. Further, one can notice the fact that the maximum deformation in the shell increases with the increase in the strain rate, but variation in magnitude is insignificant. Because the variation in the peak force is much less under different cases considered, the response of the column member of the semi-submersible is not affected by the change in strain rate.

4.6 IMPACT RESPONSE IN ARCTIC REGION

Offshore structures face a variety of challenges from the environmental and accidental loads when deployed in the Arctic region. Prevailing low temperatures induce additional threats by affecting the performance of materials. The lowest temperature in the Arctic islands and continental regions during winter has been recorded at −60°C. Structural steel used in the construction of offshore platforms suffers reduced toughness at such a low temperature. It, in turn, affects the performance of the structure. The reduced temperature in the Arctic region affects the

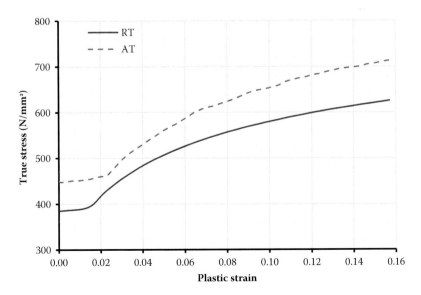

FIGURE 4.38 True stress-plastic strain curves for DH36.

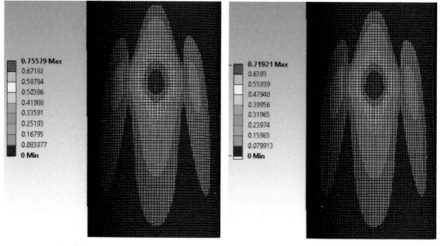

Ambient temperature   Arctic temperature

FIGURE 4.39 Damage profile of column member under Arctic temperature.

material properties significantly. At low temperature, tensile strength and yield strength of materials increases and toughness decreases. The decrease in toughness significantly affects the performance of the structures. Thus, the major problems that arise in steel are brittle fracture and corrosion. The steel material for application in the Arctic environment must satisfy the fracture toughness requirements at temperatures between –40°C to –60°C. The toughness can be increased by decreasing the grain size of steel and the addition of magnesium, copper, chrome, and nickel (Jumppanen, 1984). Thus, the ice environment plays a major role in the design and operation of offshore platforms deployed in the Arctic region.

Column members of the semi-submersible are modeled as a stiffened, cylindrical shell, and the analysis is carried out under lateral impact forces in the Arctic region. The cylindrical shell is modeled using polar class, high-tensile steel of grade DH36, based on the material properties available in the literature. The mechanical properties of DH36 steel at RT and –60°C at 0.001/s strain rate are listed in Table 4.9 (Kim et al., 2016). It is evident that the yield strength increases with the decrease in temperature. True stress-strain curves, representing the basic plastic flow characteristics of the material at room temperature and the Arctic temperature, are shown in Figure 4.38. They are used to define the material properties to predict the deformation characteristics.

From the engineering stress-strain values, the true stress-strain curves are obtained by the following equations:

$$\sigma_t = \sigma_{eng}(1 + \varepsilon_{eng}) \tag{4.17}$$

$$\varepsilon_t = \ln(1 + \varepsilon_{eng}) \tag{4.18}$$

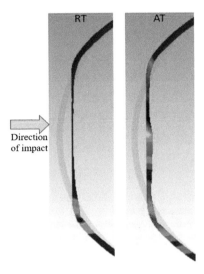

FIGURE 4.40   Damage profile of ring stiffener under Arctic temperature.

TABLE 4.10
Impact analysis under Arctic temperature

Temperature	Peak force (MN)	Maximum shell deformation (m)	Maximum energy absorbed (MJ)
Ambient temperature	8.809	0.754	6.278
Arctic temperature	8.793	0.746	6.278

where σ_t is the true stress, σ_{eng} is the engineering stress, ε_t is the true strain, and ε_{eng} is the engineering strain. The material is modeled as a multi-linear, isotropic hardening model with Von-Mises failure criteria; the effect of strain hardening is not considered.

To assess the impact behavior, an impact analysis is carried out at both the ambient and Arctic temperatures at an impact velocity of 4.0 m/s. A rectangular indenter, placed at 5.0 m above the Mean Sea Level and 9 m below the top end of the column member, is considered. Figure 4.37 shows the damage profile of the column member under Arctic temperature; the damage profile at ambient temperature is also showed for comparison. It is observed that the deformation pattern is similar under both the temperature conditions. The column member, treated as a cylindrical shell, undergoes the maximum deformation along the direction of impact in the ambient temperature. The ring stiffener R3 gets flattened at the impact location and undergoes twisting as shown in Figure 4.38. The maximum strain in the ring stiffeners (R2, R4) is only about 12% of that of R3. It confirms that the ring

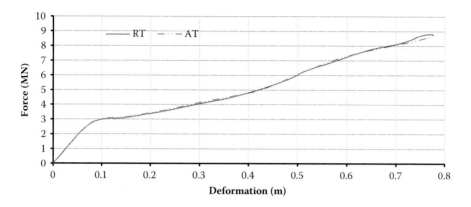

FIGURE 4.41   Force-deformation curve of column member under Arctic temperature.

stiffeners constrain the spread of damage to the adjacent bays, which is an important aspect of design (Figures 4.39 and 4.40).

The results of impact analysis under Arctic temperature are summarized in Table 4.10. Though the impact occurs at the location of ring stiffener R3, the maximum deformation along the impact direction is observed only in the stringer stiffeners. The peak impact force developed during collision is found to be maximum in the ambient conditions due to the reduced yield strength of the material. Force deformation curves at ambient and Arctic temperatures are shown in Figure 4.41; the area under this curve is the measure of total energy absorbed by the column member.

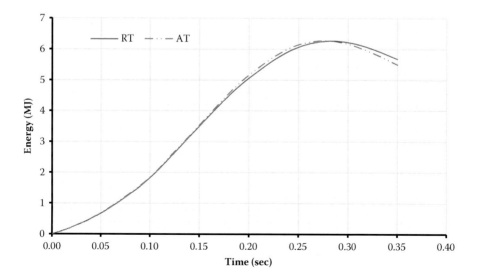

FIGURE 4.42   Energy absorbed by column member.

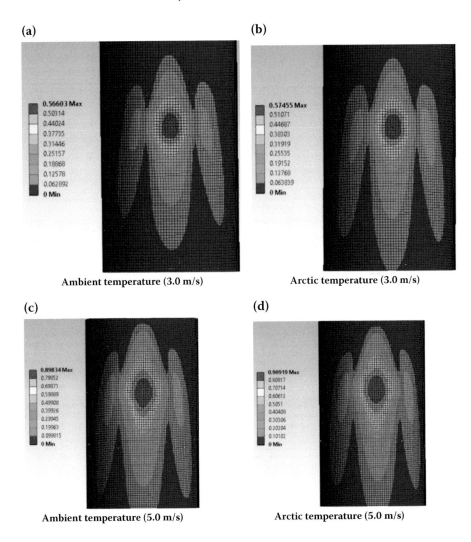

(a) Ambient temperature (3.0 m/s)

(b) Arctic temperature (3.0 m/s)

(c) Ambient temperature (5.0 m/s)

(d) Arctic temperature (5.0 m/s)

FIGURE 4.43 Damage profile of column member for various indenter velocities at Arctic temperature.

In Figure 4.41, flattening of the curve at 0.1 m indicates torsional buckling of the stiffeners. Under both the temperature conditions, force-deformation curves are almost identical at the initial stages, showing a good resistance to deformation. But the ultimate resistance to deformation is reduced at the Arctic temperature irrespective of the increased stiffness in the cylindrical shell. It can be attributed to the reduction in fracture toughness of the material. Energy absorbed by the member at different temperatures is shown in Figure 4.42. It is seen from the figure that the column member absorbs the maximum energy at a comparatively shorter impact duration at the Arctic temperature. From the longitudinal deformation pattern, it is observed that the indentation is not extended up to the upper end of the member.

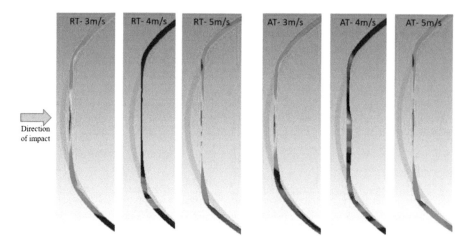

FIGURE 4.44 Deformation of ring stiffener R3 for various indenter velocities.

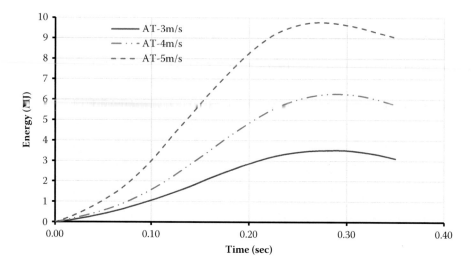

FIGURE 4.45 Energy absorbed by column member at Arctic temperature under different indenter velocities.

TABLE 4.11
Impact analysis under different impact velocities at Arctic temperature

Temperature	Impact velocity (m/s)	Peak force (MN)	Maximum shell deformation (m)	Maximum energy absorbed (MJ)
Ambient	3	6.581	0.584	3.528
temperature	5	11.211	0.931	9.812
Arctic temperature	3	6.676	0.575	3.53
	5	11.473	0.909	9.81

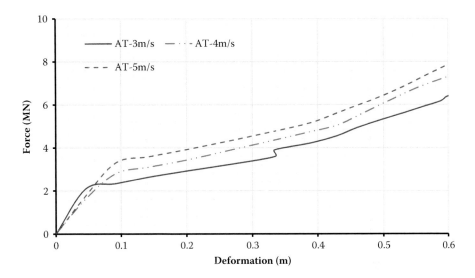

FIGURE 4.46   Force-deformation at Arctic temperature under different indenter velocities.

TABLE 4.12
Impact test results for different indenter sizes at Arctic temperature

Temperature	b/R ratio	Peak force (MN)	Maximum deformation (m)	Maximum energy absorbed (MJ)
Ambient temperature	0.133	11.763	0.64	6.267
	0.40	7.535	0.901	6.158
Arctic temperature	0.133	11.859	0.631	6.277
	0.40	7.77	0.907	6.227

Plastic strain is also not observed near the Mean Sea Level, which reduces the danger of flooding of compartments during impact. It can be stated that the polar-class marine steel, developed for low-temperature applications, is highly capable of alleviating the loads effectively. It also confirms the importance of proper material selection concerning to the environmental conditions of operation.

4.6.1 INDENTER VELOCITY

The effect of the impact velocity on the column member of a semi-submersible is assessed by varying the velocity from 3–5 m/s, at both ambient and the Arctic temperature. As seen from the damage profile of the column member in Figure 4.43, the impact velocity increases with the area of contact between the indenter and

(a)   (b)

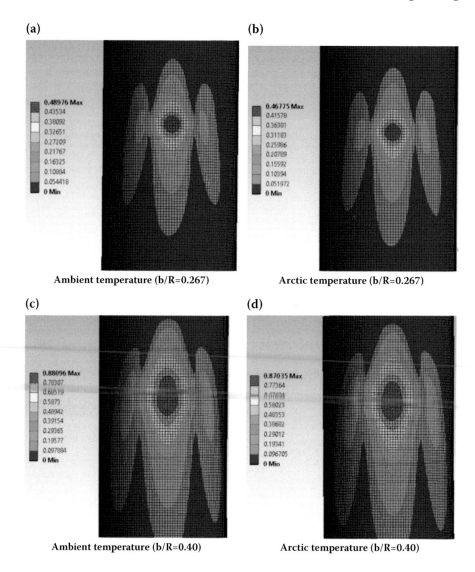

Ambient temperature (b/R=0.267)   Arctic temperature (b/R=0.267)

(c)   (d)

Ambient temperature (b/R=0.40)   Arctic temperature (b/R=0.40)

FIGURE 4.47   Damage profile of column member for various indenter sizes in Arctic temperature.

member. The deformation in the ring stiffener R3 also increases with an increase in the velocity (Figure 4.44). The results of the impact analysis are summarized in Table 4.11. The table shows that the variation in the deformation of the cylindrical shell under different velocities is less than 5%. At low velocities, the impact force is resisted by stringer stiffeners, and maximum deformation is observed only in them. With the increase in the impact velocity, the ring and stringer stiffeners together resist the higher impact force by increasing the stiffness of the cylindrical shell. The energy absorbed by the cylindrical shell increases with an increase in impact

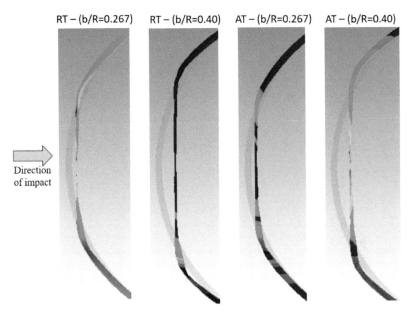

FIGURE 4.48   Deformation of ring stiffener R3 for various indenter sizes.

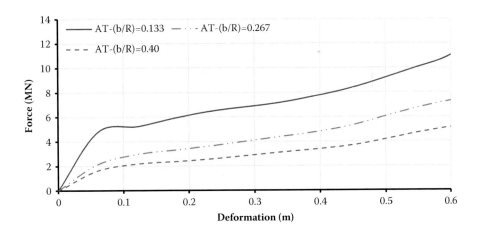

FIGURE 4.49   Force-deformation at Arctic temperature for various indenter sizes.

velocity, as shown in Figure 4.45. It can be seen that the peak energy is attained at a shorter duration when the impact velocity is high. The force-deformation characteristics of the column member do not show significant variations at both the ambient and Arctic temperatures, as observed from Figure 4.46. Material behaves effectively at both the temperature conditions due to its suitable mechanical properties. At 5.0 m/s impact velocity, the slope change in the force-deformation curve occurs at two

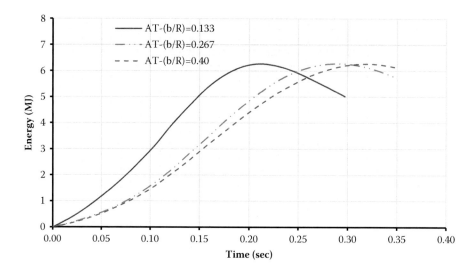

FIGURE 4.50 Energy absorbed by the column at Arctic temperature for various indenter size.

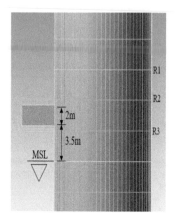

FIGURE 4.51 Indenter location at mid-bay between ring stiffeners.

stages: First, flattening of the curve occurs due to torsional buckling of the stringer stiffeners; second, the slope increases due to an increase in the resistance to impact force offered by the combined action of the ring and stringer stiffeners. Thus, it can be stated that the orthogonally stiffened cylindrical shells are highly advantageous in resisting the impact force at higher velocities and are common design configurations for column members of semi-submersibles.

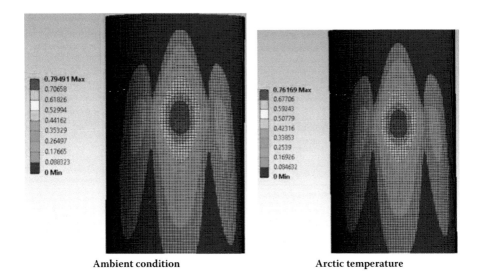

Ambient condition   Arctic temperature

FIGURE 4.52   Damage profile of column under Arctic temperature.

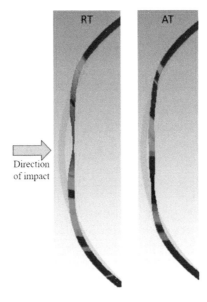

Direction
of impact

FIGURE 4.53   Deformation of ring stiffener R2.

4.6.2 INDENTER SIZE

Numerical analyses are carried out by varying the indenter depth in the Arctic temperature; results are summarized in Table 4.12. It is observed that the peak force, which is developed during collision, decreases with the increase in the (b/R) ratio. The depth of the

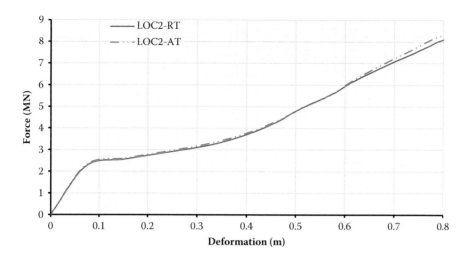

FIGURE 4.54 Force-deformation curve under mid-bay collision.

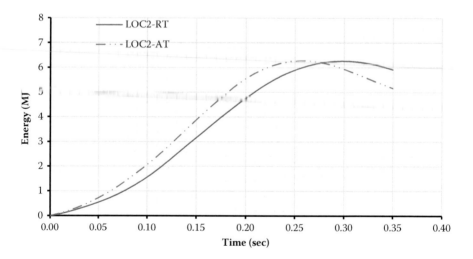

FIGURE 4.55 Energy absorbed by buoyant leg under mid-bay collision.

indenter increases with the contact area, which causes an increased depth of local in-dentation. Comparatively higher deformation is observed at the Arctic temperature.

Figure 4.47 shows the damage profile of the column member at Arctic temperature for various indenter sizes. Figure 4.48 shows the deformation of ring stiffener R3 for various indenter sizes at the Arctic temperature. It is observed from the figures that an increase in the size of the indenter increases the area of local indentation. It subse-quently increases the corresponding strain at the impact location. The extent of da-mage in the circumferential and longitudinal direction also increases with an increase in the (b/R) ratio. The pattern of damage in all cases is found to be similar. The maximum strain is seen only within the adjacent bays of the impact location due to the presence of ring stiffeners. At lower indenter depths, the maximum deformation is observed only in ring stiffener R3. But with an increase in indenter depth, ring stiffeners R2 and R4 also undergo deformation.

The force-deformation curves under different cases are shown in Figure 4.49 at the Arctic temperature. It is seen from the plot that at smaller (b/R) ratios, the slope of the force-deformation curve is high compared to the other two cases. Flattening of the curve is seen at a (b/R) ratio of 0.133, which occurs due to torsional buckling of the stiffeners. The initial slope of the curve decreases with an increase in the (b/R) ratio. The impact energies absorbed by the column member under different cases are shown in Figure 4.50. It is seen from the plot that the maximum energy is attained at a shorter duration for a smaller (b/R) ratio. A marginal variation in the energy-time history is absorbed at ambient temperature and the Arctic temperature. At the Arctic temperature, the variation in the absorbed energy is less at higher (b/R) ratios.

4.6.3 MID-BAY COLLISION

Impact analysis is carried out by varying the indenter location as shown in Figure 4.51 in the Arctic temperature. The center of impact of the indenter of 2.0 m depth is located at mid-bay between ring stiffeners R2 and R3. The indenter velocity and energy absorbed by the column vary with the change in the impact location and temperature. The deformed shape of the column at Arctic temperature is shown in Figure 4.52. A comparatively higher area of indentation is observed at room temperature. The deformation of ring stiffener R2 is shown in Figure 4.53. Stringers resist the impact load by beam action and collapse due to the formation of a plastic hinge. Because of the change in the circumferential strain and increased deformation of the shell, the ring stiffener undergoes tilting. Both the ring stiffeners R2 and R3 undergo highly concentrated strains at the ends of the flattened section. This shows that the ring stiffeners play a major role in resisting the impact force at both mid-bay collision and the ring-frame collision.

Force-deformation curves for mid-bay impact for both the ambient condition and Arctic temperature are shown in Figure 4.54. It is seen that the maximum impact force is developed under the Arctic temperature when the indenter collides with the cylindrical shell at mid-bay. However, variation in the force-deformation is observed only after 0.6 m deformation along the direction of impact. Stringers undergo the maximum deformation compared to that of the ring stiffeners. Stringers undergo yielding, which is followed by the yielding of the ring frames. This is also evident from the flattening of the force-deformation curves at both the temperature conditions. Under this colliding condition, ring stiffeners R2 and R3 undergo almost equal deformation. The energy absorbed by the column is shown in Figure 4.55. Column members absorb the maximum energy under the Arctic temperature. However, the magnitude of the peak energy remains unaltered.

5 Pipe-Laying Barges

5.1 INTRODUCTION

Offshore structures refer to all those structures operating in the open sea, which are directly inaccessible from the dry land. These structures are capable of functioning as stand-alone systems in various ocean climates. They differ from one another depending upon the water depth in which they are deployed. Initially, exploration and production operations are carried out only in shallow water depths, which are recently extended to deep and ultra-deep waters. As the operation of vessels moved from shallow to deep waters, many floating structures were commissioned for various purposes. The main functions are oil and gas exploration, production, and storage. Further, these structures are also extended toward floating airports, bridges, breakwaters, offshore wind and solar energy farms, and floating docks. Floating offshore structures are categorized as neutrally buoyant and positively buoyant.

5.1.1 FLOATING PRODUCTION UNITS

Floating production units (FPUs) are oil production facilities for minor fields. Figure 5.1 shows the schematic figure of a typical production unit, which is moored to the ground. It does not possess any storage capabilities. Oil produced from the well is offloaded onto a Floating, Storage, and Offloading (FSO) unit. There are typically two types of FPUs, namely, tension leg platforms and semi-submersibles. In both cases, a storage tanker has to be arranged nearby the FPU to store the explored oil and gas. These facilities are connected using pipelines to transfer oil from the FPU to the tanker. Pipelines should be flexible and durable to withstand the wave-induced motion of the floaters.

5.2 SPECIAL PURPOSE VESSELS

5.2.1 PIPE-LAYING VESSELS

Pipelines are used to transport oil from the production platform to the shuttle vessel. Because during laying of the pipelines, they will be under bending stress of higher magnitude, the probability of their failure during the laying operation is also higher. One of the simple and conventional techniques used to minimize such failure is to maintain a high axial tension in the pipes during the laying operation. This is similar to inducing high pre-tension in the pre-stressed concrete members. Axial tension, if maintained during the laying operation, will prevent the pipeline from buckling failure under its self-weight. Specially designed tensioners provide the required tension for the safe deployment of the pipe into the sea. It is usually positioned at a location near the release point of the pipeline from the vessel. The design of the tensioners is critical, as it should not damage the outer covering of the pipe. Further, it should provide sufficient

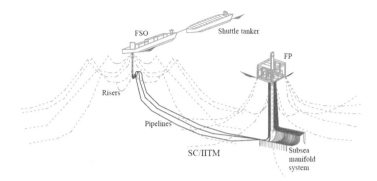

FIGURE 5.1 Floating Production Unit with external storage.

grip to the pipeline as the vessel moves forward to ease its release smoothly. On one side of the tensioners, additional pipes are welded on to the existing pipeline, while on the other side, freely spanning pipe is supported by stingers.

During the laying operation, a vessel, dedicated for this purpose, is necessary. This vessel should house the pipeline reels (or hoses), which will be laid as the vessel moves in the forward direction with a gentle velocity. Such vessels are called pipe-laying barges. These barges are designed similarly to a semi-submersible with a ship-shaped hull. One of the main advantages of possessing a semi-submersible design is that this design offers excellent stability against waves. It, in turn, enhances both the safety and comfortability of the pipe-laying operation. Ship-shaped vessels provide improved load capacity and high transit velocity. In the course of pipe laying, the vessel is continuously repositioned with the help of anchors or dynamic positioning systems (DPS). Figure 5.2 shows the schematic drawing of a pipe-laying vessel.

Pipe-laying vessels are analyzed for critical wave orientation, which is usually the bow-quartering condition. It accounts for the critical combination for both the vessel and the pipe motion. A pipe-laying operation by a moored vessel is mainly dependent on a few factors: mooring stiffness, forces acting on the mooring lines, risk of handling anchors, and motion in the dominant wave-frequency. In addition, compliant motion of the vessel, holding capacity of the anchor, forces acting on the stinger-vessel connectors, and bending and buckling stresses acting on the pipeline also influence the pipe-laying operation.

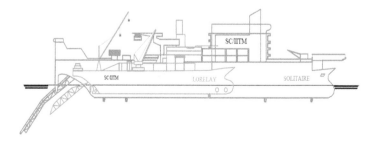

FIGURE 5.2 Pipe-laying vessel.

5.2.2 DRILLING VESSELS

When a potential site is identified for oil and gas exploration using seismic surveys, a well is bored to investigate these findings. Barges, Jack-up rigs, semi-submersibles, and drillships are deployed to carry out the exploratory drilling. Figure 5.3 shows the schematic views of different drilling vessels. A floating, drilling vessel is preferred to operate in high sea states due to its adaptability for the rough sea. As discussed in detail, the workability of semi-submersibles is quite excellent due to their low motion in the vertical plane. At locations where significant motion is expected in the vertical plane, a heave compensator is used. It helps handle the drill string of the drilling tower. It maintains the drill string at the essential tension required during the heaving of the vessel (Huang and Dareing, 1968). By these additional forces, loads acting on the drill bit are reduced under vertical motions. Apart from the drill strings, risers are also sustained under tension using the riser-tensioning systems (Rustad, 2008).

Motion in the horizontal place also plays an essential role in the operation of such vessels. It is expected that the top end of the drill string should be moved beyond the equilibrium point by about 5% of the water depth. It can be achieved by a carefully designed mooring system.

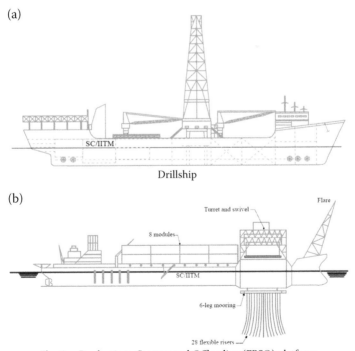

(a)

Drillship

(b)

Floating Productioon Storage and Offloading (FPSO) platform

FIGURE 5.3   Drilling vessels used for oil and gas exploration.

5.3 SUPPORT VESSELS

Apart from the regular vessels that are used for storage and transportation of oil, there are floating structures that assist the regular vessels operating side by side. These vessels are known as support vessels, or Offshore Support Vessels (OSVs). Some of the OSVs are a seismic survey vessel, platform supply vessel, crane vessels, well-maintenance vessels, ROV support vessels, and diving-support vessels.

5.3.1 SEISMIC AND SURVEY VESSELS

Oil and gas reserves in a particular location need to be ascertained before the exploratory drilling. During the initial days of offshore drilling, these activities were carried out using existing vessels. In the later stages, where oil and gas locations are extended to rough environments, specialized vessels are being deployed. These vessels are commonly known as Seismic and Survey vessels (Figure 5.4). These vessels carry sophisticated equipment, which gathers data for the system analysis.

5.3.2 WELL-MAINTENANCE VESSELS

Oil wells needed to be maintained throughout their service life for better yield. It is vital that these wells are maintained periodically, and this is done with the help of dedicated vessels known as well-maintenance vessels. These floating structures are equipped with sophisticated instruments that specifically work toward achieving this purpose. Tools are cautiously lowered into the well for inspection and repair works. These vessels are position-restrained with DPS.

5.3.3 DIVING-SUPPORT VESSELS

These vessels are used for underwater applications like assessment and maintenance of compliant platforms, pipelines, cables, and other subsea systems. They are fitted with a moon-pool for providing access to the divers, remote-operated vessels (ROV), and various tools to lead through. It is important to note that these vessels

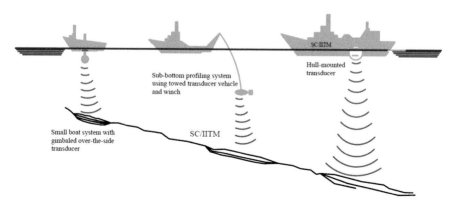

FIGURE 5.4 Seismic survey vessel systems.

need to be precisely positioned as the margin for the operation is highly limited. They are invariably position-restrained using DPS.

5.3.4 CRANE VESSELS

Marine vessel operations require a lot of transfer of equipment and other goods, to and from the vessel, continuously. These operations are performed using heavy-duty cranes, which are housed on crane vessels. Such cranes will not be housed permanently on the production platform due to many reasons (Kim and Hong, 2009). Apart from imposing heave dead load on the platform, they reduce the free-working space on the top side (Ngo and Hong, 2010; 2012). Further, there are no permanent requirements for such cranes. Hence, it is always a common practice to hire the crane vessels, as and when required. It is also convenient to deploy the crane vessels as the exploration, production, and goods handling can take place side-by-side. It radically reduces the design complexities of the offshore platforms. Besides, it increases the lifting capacity as crane vessels are dedicated ones. Figure 5.5 shows images of typical crane vessels used in offshore drilling operations.

5.4 TRANSPORT VESSELS

Conventional transportation vessels include supply vessels, tugs, shuttle tankers, transfer vessels, launch barges, and heavy-lift transport vessels. They are useful support-systems during oil and gas exploration. It is interesting to note that as offshore oil drilling and production is executed with a limited crew, additional support offered by such vessels is essential for the smooth and safe execution of work.

5.4.1 SHUTTLE TANKERS

Produced oil from the ocean needs to be transferred to the shore by some means. This is due to the limited capacity to store oil in the production platform itself. Further, the

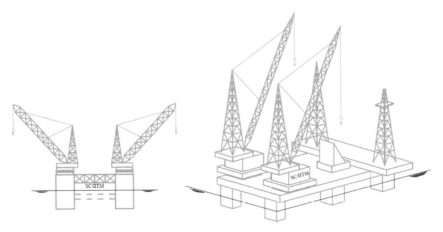

FIGURE 5.5 Crane vessels.

storage of oil also causes a potential threat of fire accident, as hydrocarbons are highly explosive. Transport of oil to the shore is usually done through pipelines. But, under the current scenario of deep-water drilling, offshore platforms are positioned quite far away from the coast. Providing pipelines imposes a high risk of failure due to longer segments, apart from being very expensive. So an alternate practice is to transport the produced oil through tankers called Shuttle tankers (Figure 5.6). These are modified forms of sea-going vessels that can be utilized for temporary storage and transfer of oil from site to the shore or other facilities. Production platforms like Floating, Production, Storage, and Offloading (FPSO) work side by side with a shuttle tanker to transfer oil. Shuttle tankers will be positioned near the facility, either with a DPS or a mooring system; sometimes, both are used in tandem.

5.4.2 Heavy-Lift Transport Vessels

These vessels are high-capacity cargo barges, which are self-propelled or towed to the site. The cargo is being loaded onto these barges by a roll-on process from the shore. Alternatively, the transfer also takes place by a float-on operation by submerging the deck. Figure 5.7 shows a typical heavy-lift transportation vessel deployed in the Gulf of Mexico (GoM). The figure shows the transportation vessel is

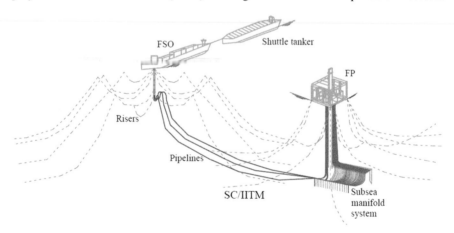

FIGURE 5.6   Shuttle tanker and FPSO.

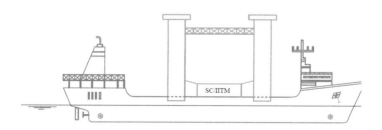

FIGURE 5.7   Heavy lift float-on float-off vessel.

housing a TLP and sailing. Self-propelled transportation vessels, apart from carrying massive marine structures and equipment, can also drive at a significant speed. It reduces the towing time of offshore floating structures, significantly.

5.4.3 Launch Barges

These are enormous vessels or barge structures with a flat top, which are used to transport long-span offshore structural members like jackets and other peripheral components. While launching a jacket member off the barge, fasteners are pulled off and the jacket is guided through the launch ways. The barge is constrained to remain in position during the whole process. A barge is ballasted to an extent for easy and convenient sliding of the jacket through the guideways. The jacket member leaves the barge while passing over a rocker beam. In some cases, the jacket is lifted off the barge with the help of cranes.

5.4.4 Tugs and Supply Vessels

Tugboats are used to tow floating offshore structures from one location to another. Typically, they act as a permanent associate in the drilling operation. Anchor-handling tugs are specialized tugboats that are exclusively used for changing the anchor locations during the operation of a pipe-laying vessel. Supply vessels are used to supply different tools and equipment required for production and other associated facilities, namely, drilling strings, drill bits, pipelines, and risers. As there will be a continuous demand for such accessories from the production facility, aft of the supply such vessels are kept as ample open space to facilitate the storage and movement of equipment, in and out of the supply vessel. Figures 5.8 and 5.9 show a schematic view of a tug boat and supply vessel, respectively.

5.4.5 Transfer Vessels

Transfer of the crew to and from the offshore site is usually done using helicopters. Transfer vessels are also deployed for this purpose. They are also called Crew boats.

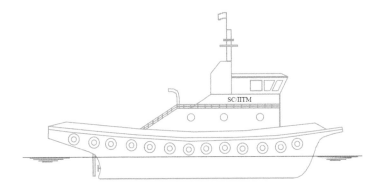

FIGURE 5.8 Tugboat.

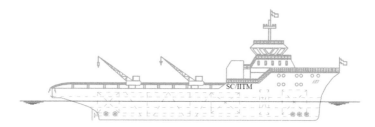

FIGURE 5.9 Supply vessel.

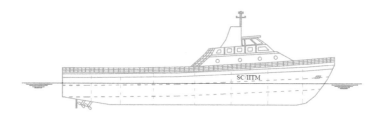

FIGURE 5.10 Crew boat.

(a)   (b)

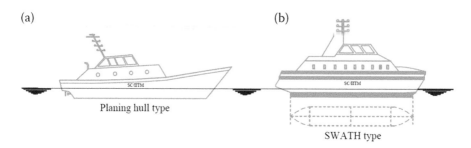

Planing hull type

SWATH type

FIGURE 5.11 Pilot vessel.

Other forms of vessels used for the transfer of personnel are pilot vessels, which are operated at a much slower speed. They are designed for short-distance travel. Figure 5.10 shows a schematic view of a Crew boat. Figure 5.11 shows the view of a Pilot vessel, used for patrolling and transfer of personnel in an emergency.

5.5 MOORING SYSTEMS

Mooring lines must be provided for floating offshore structures to position-restrain during their operation. Mooring systems consist of cables, which are free-hanging from the platform, and are connected to the seabed through anchors or piles. These anchors are placed at a desirable distance away from the platform to avoid inter-ference during drilling. It is a common practice to adopt a symmetric mooring layout. Chain and wire ropes are usually used as mooring lines. They are used independently or in a combination of segments for carrying out operations in specific conditions. Each line provides restoration for the structure, depending on the displacement of the vessel. Accordingly, a change in the geometric shape of the mooring line is experienced while the line is either lifted off or laid onto the ground. Under both circumstances, mooring lines develop nonlinear restoring forces to compensate for the station keeping of the vessel or the barge. The restoring force generated by the mooring lines depends on the offset of the vessel in the horizontal plane. As discussed in the earlier chapters, floating offshore structures and large barges are highly flexible in the horizontal plane, but remain stiff in the vertical plane. The stiffness developed to resist the vertical motion is the result of the positive buoyancy of the vessel or the platform. Hence, motion caused by the waves is significantly less than that of the vessel motion. An equivalent restoring stiffness of the mooring lines will control the position-keeping. Low-frequency drift excitations can significantly amplify the vessel motion in the horizontal plane, which thereby increases the tension on the mooring lines significantly.

As the depth of operation increases, a simple chain mooring will not be sufficient and economical due to its high suspended weight. Hence, synthetic fiber ropes are used for taut-mooring. There are many advantages of using fiber ropes. Namely, less self-weight, reduced tension at the fairlead, and controlled length of the mooring line as well. Mooring lines offer sufficient compliancy along with adequate restoration to excessive offsets. It prevents damage to the drilling units and other equipment. Recent studies have shown that moorings are used in conjunction with the thruster-assisted DPS. A spread, catenary mooring is used for vessels where the heading of the vessel is significantly high. Mooring lines should possess a few characteristics, such as positioning capability, easy installation, offset restrictions, and serviceability. Figure 5.12 shows various environmental forces that act on a moored vessel.

Safety is the most critical aspect of an offshore structure. It becomes even more critical while the platform is in operation. Even wave forces with second-order effects

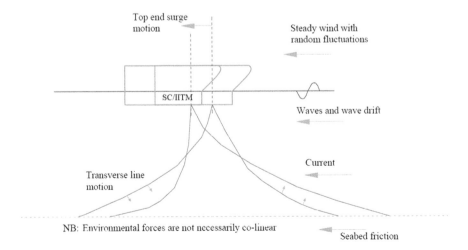

FIGURE 5.12   Environmental forces acting on a moored vessel.

can cause a severe impact on the platform or the vessel and subsea systems. With time, the offshore entities need to be stable in their operation and production period even if the environmental conditions, like wind, wave, or current, change (Tran et al., 2016). To safeguard the operation of the platform, motion-control mechanisms are implemented (Karperaki et al., 2016). The re-centering ability of compliant offshore structures can be improved by ample buoyancy, tethers, or a combination of both (Adrezin, 1996). To maintain the design-draft during the drilling operation, a station-keeping system is necessary. The choice of these systems is determined by the service requirements and characteristics along with the depth of water at the operational condition and sea state.

Further, the choice of a mooring system is governed by a few parameters, namely, overall weight and length, and the axial stiffness capacity. If the system requires a substantial impact from the axial rigidity of the mooring, consideration is given to a taut-mooring. At the same time, the submersion weight of the line ensures reinstatement in a catenary system (Felix and Mercier, 2016).

One of the commonly deployed and preferred mooring systems is the spread mooring system. This system is well-suited for shallow and intermediate water depths, up to 500 meters. With the increase in depth, the feasibility of its use is restrained due to the increased economic, technical, and operational difficulties. Therefore, for deep-water operations, taut moorings or DPS are preferred. Spread mooring systems offer longitudinal restoration to the platform or the vessel, as the case may be with a minimum contribution to the vertical restoration. Besides, it also contributes to the overall damping of the system. In large structures, its contribution to the low-frequency motion is quite useful but doesn't restrain the first-order motion of the vessel (Oppenheim and Fletter, 1991).

Moorings of complaint offshore platforms and other support vessels are essential for safe operations while ensuring the desired station-keeping characteristics and re-centering capabilities (Chandrasekaran, 2015a,b). FORM-dominated offshore compliant structures are designed to inherit compliancy in the horizontal plane through mooring lines and tethers. See, for example, tension leg platforms, Triceratops, oil tankers, semi-submersibles, drillships, and pipe-laying barges. A massive displacement is permitted in the horizontal plane, as a desirable feature of the design (Chandrasekaran et al., 2016b; Chandarsekaran and Sihhas, 2016; Chandrasekaran, 2014). Careful selection of geometric configuration and material for mooring lines, clump weights, buoys, and anchors are crucial for safety and survival in different sea states (Khan and Ansari, 1986; Chandrasekaran, 2015b).

Mooring systems show different degrees of complexities based on the functional requirements of the marine vessels (Pedersen, 1975; Ellermann et al., 2002; Ellermann, 2005; Ellermann and Kruizer, 2000). The low-frequency drifting force can excite higher period surge and sway motions with large amplitudes on a floating platform or a vessel, which is position-restrained with anchor lines (Verhagen, 1970). Marine structures have a single point mooring (SPM) system, which is typical for both the buoys and the tankers. To facilitate a convenient operation, spread mooring systems are deployed in drilling platforms and semi-submersibles (Clauss et al., 2000; 2002). Complexities in numerical analyses increase with the increase in number of cables in the mooring system (Garza and Bernitsas, 1996; Garza et al., 2000). These factors include the number, orientation, and pre-tension of the mooring lines.

Among different properties of the mooring lines, stiffness is the most vital property (Oppenheim and Fletter, 1991). It is used to represent the general and physical characteristics of a station-keeping system. Further, stiffness value can be an indicator of water depth, the material used for the mooring line, its pre-tension, and mooring line patterns. It also governs a few practical applications, namely, anchor-positioning, length of the line, directionality, the orientation of the spread system, nonlinearities, and sensitivity to any modifications. In some coastal regions, the mooring of a vessel becomes very difficult due to its geographical and weather conditions. These vessels can exhibit significant movement and eventually lead to the failure of the mooring systems, even under normal weather conditions (Van Oortmerssen, 2008).

In specific terms, stability and response of the large vessels, moored with spread mooring system, become essential. Even though numerical techniques have limitations in the transcription of this physical problem, they are suitable for multi-degree-of-freedom systems due to their flexibility in handling the problem. Various researchers have successfully used polynomial approximation to reduce the time and numerical efforts of simulating mooring problems (Ellermann, 2005; Ellermann and Kreuzer, 2000; Ellermann et al., 2002; Umar and Datta, 2003). Numerical simulations are used to explain the nonlinearity in the response of a moored buoy; results are comparable with that of the experimental investigations (Ellerman and Kreuzer, 2000). Complexities arising from geometric nonlinearities of mooring lines are successfully handled as white noise with an appropriate probability density function (Roberts, 1981).

5.6 SHALLOW WATER EFFECTS

Water bodies in which the vertical boundaries are closer to the floating vessel are known as shallow water. The reduced depth of water column induces resistance, which, in turn, makes the vessel motion tougher. The pressure developed as the ship motion increases will result in waves of larger amplitude compared to the deep-sea state. In addition, changes in the velocity stream, past the vessel, will add to the resistance offered by it (Saha et al., 2004). Fluid-flow characteristics vary in different water depths. In the case of deep water, the bottom of the sea is considered flat; any error incurred due to this is assumed to be insignificant. But while considering the shallow water condition, this assumption doesn't hold true. The dynamic effect of the fluid is expected to act throughout the entire depth of water. Propagation and refraction of the wave-field largely depend on the structure of the seafloor (Teigen, 2005).

The water particle kinematics are primarily influenced by the seabed topology. As the water depth changes from the deep water to a shallow water condition, the Shoaling effect is encountered (Dean and Dalrymple, 1991). The deep-water condition is said to be satisfied when the depth of the water column exceeds half of the wavelength of the most extended wave in the ensemble. In such conditions, the topology of the seafloor does not significantly influence the water particle kinematics (Harito, 2007). While considering the shallow water wave region, the Airy wave is considered to be a regular, linear wave. Stokes higher-order wave theory,

Solitary wave theory, and cnoidal wave theory are grouped as regular, nonlinear waves. The realistic ocean waves will be uneven, which can be reflected as the superposition of different types of regular waves. The irregularity of these waves is well-described with the help of a wave spectrum (Li, 2008). The boundary element problem is resolved using shallow water Greens function to acquire the response amplitude operator (RAO) of the structure using the Laplace equation. The nonlinearity and severity of the shallow water conditions cannot be effectively and directly represented using the JONSWAP spectrum. It needs to be modified with the shallow water alteration factor (Bouws et al., 1985). Shallow waters provide more challenges to the numerical modeling of mooring line analyses due to high nonlinearities that arise from the lower water depth (Xiao et al., 2014). Despite less wave frequency motion in shallow water, vessel motion shows a broader bandwidth of frequencies (Li et al., 2003). Time-domain approaches are employed to deal with the analysis of the transient occurrences due to the steep waves and effects of high nonlinearity. It comprises direct integration techniques (Watanabe et al., 1998) and Fourier transform schemes (Kashiwagi et al., 2000; Montiel et al., 2012).

Various approaches can be applied to determine wave forces based on different flow regimes that are defined around the barge. Morison's equation is useful where the geometric dimensions are smaller than the wavelength. In simple terms, the drag component of the force will become dominant. Characteristic dimensions of the barge tend to affect the flow regime around it. Based on the dimensions, the region shall experience scattering or reflection of the waves, outward from the barge. In such cases, the diffraction theory is applied to compute the wave forces. Diffraction theory is applicable when the ratio of characteristic dimension to a wavelength of an incident wave exceeds 0.2. Underlying assumptions taken into consideration are as follows: (i) fluid is considered to be ideal; (ii) flow is irrotational; (iii) the amplitude of the incident wave is considered to be small, and (iv) the linearized Bernoulli equation is used to acquire the fluid pressure.

5.7 SEA STATE MODELING

Most important of all, the forces acting on the barge are wave forces, particularly those generated by wind. Wind, which causes these waves, is locally generated by intense storms, originating far away from the shore. Undulations formed due to the wind on the surface of the sea are restored to equilibrium position by the action of gravity. Hence, such waves are mainly known as wind-generated gravity waves. Figure 5.13 shows different wave profiles.

Another classification of ocean waves is based on the wave period. If the wave occurrence is of much less frequency, then it is called a long-period wave. Alternatively, if the wave occurrence is frequent, such waves are known as short-period waves or gravity waves. Usually, the period of gravity waves lies in the range of 4 to 30 s. These waves are generated by wind and restore themselves by the action of gravity. Progressive waves are those waves in which the profile of the wave moves forward as a function of time; water particles move with respect to time. When the water particles show vertical oscillations about the wave's mean position, the wave is known as a standing wave or clapotis wave. Small amplitude waves are those whose

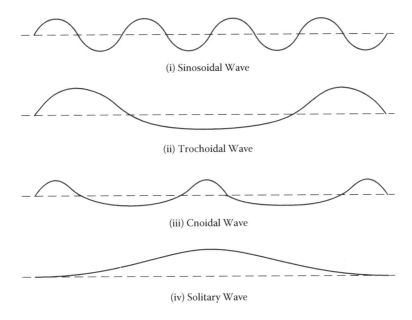

(i) Sinosoidal Wave

(ii) Trochoidal Wave

(iii) Cnoidal Wave

(iv) Solitary Wave

FIGURE 5.13   Wave profiles.

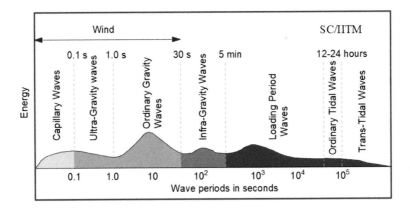

FIGURE 5.14   Wave energy content.

wave steepness is smaller than 0.02; otherwise, they are called finite-amplitude waves. Figure 5.14 shows the energy content of different types of waves, with their variations in the period.

5.8 PIPE-LAYING BARGE

Figure 5.15 shows the schematic view of a typical, moored vessel, which is under consideration. Table 5.1 shows the geometric properties of the vessel. Figure 5.16 shows the schematic view of the pipe-laying barge with an eight-point mooring configuration.

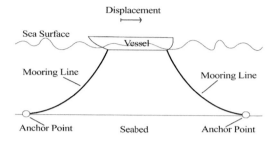

FIGURE 5.15   Schematic view of a moored vessel.

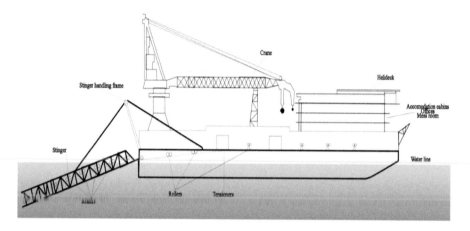

FIGURE 5.16   Schematic view of the pipe-laying barge.

TABLE 5.1
Geometric properties of the barge

Description	Value
Length (L)	79.40 m
Breadth (B)	35.35 m
Operating draft (D_r)	2.50 to 3.0 m
Depth	4.27 m
Mass (M)	$6.77 * 10^3$ ton
Center of gravity (CG)	(38.64,0.00,4.78) m
Radius of gyration about X-axis (r_{xx})	10.10 m
Radius of gyration about Y-axis (r_{yy})	27.10 m
Radius of gyration about Z-axis (r_{zz})	25.60 m
Lightship load	3500 ton
Main Crane	380 ton
Deck load	2800 ton
Generator sets	2 × 500 kW

The name and characteristic features of the vessel are masked for strategic reasons. Draft and free-board of the vessel are chosen as per the exact details. The response of the vessel is obtained for different wave heading angles in all active degrees-of-freedom. Tension variations in the mooring lines are estimated, based on which new mooring layout is proposed to improve the re-centering capability of the vessel under a moored condition. Tables 5.2 and 5.3 show the details of the sea state for 3 and 6 m water depth, respectively.

Pipe-laying barges experience fluctuating loads under the encountered waves, which accelerates significant motion in the horizontal plane. For the known sea surface elevation near the barge, forces arising due to variable submergence add nonlinearity to the response. Transformation from sea surface elevation to the exerted force on the vessel is less influenced by the sea state in comparison to that of the feedback force that arises due to the response of the barge. The force exerting on the barge also depends on the barge's position with respect to head-sea conditions. In the present study, the catenary theory is used to obtain load-excursion characteristics of the mooring line (Baltrop, 1998). Potential flow theory is used to model the vessel behavior. A few underlying assumptions used in the analysis are namely: i) fluid is inviscid, ii) fluid is incompressible, and iii) flow is irrotational.

TABLE 5.2
Details of sea state for 3 m water depth

Water depth = 3 m

Swell and Associated Parameters for Site		**Wind Sea and Associated Parameters for Site**	
Parameter/Return Period	1 year	Parameter/Return Period	1 year
Swell waves		**Wind waves**	
Maximum wave height (m)	2.34	Maximum wave height (m)	2.34
Significant wave height (m)	1.52	Significant wave height (m)	1.45
Direction (°)	30	Direction (°)	45
Period (sec)	15	Period (sec)	7.1
σ (Gaussian spectral width)	0.007	γ (JONSWAP)	1.8
Associated wind wave		**Associated swell**	
Significant wave height (m)	1.1	Significant wave height (m)	1.6
Direction (°)	45	Direction (°)	30
Period (sec)	6.3	Period (sec)	13.7
γ (JONSWAP)	1.8	σ (Gaussian spectral width)	0.007
Tide	1.2		
Wind		**Wind**	
Wind with wave, 10-min, 33′ (kt)	14	Wind with wave, 10-min, 33′ (kt)	27
Direction (°)	225	Direction (°)	225
Current (in line with wave)		**Current (in line with wave)**	
Surface speed (knot)	0.5	Surface speed (knot)	0.5
3 ft off bottom (knot)	0.3	3 ft off bottom (knot)	0.3

TABLE 5.3
Details of sea state for 6 m water depth

Water depth = 6 m

Swell and associated parameters for site		Wind sea and associated parameters for site	
Parameter/Return Period	1 year	Parameter/Return Period	1 year
Swell waves		**Wind waves**	
Maximum wave height (m)	4.5	Maximum wave height (m)	4.75
Significant wave height (m)	2.02	Significant wave height (m)	1.94
Direction (°)	30	Direction (°)	45
Period (sec)	15	Period (sec)	7.1
σ (Gaussian spectral width)	0.007	γ (JONSWAP)	1.8
Associated wind wave		**Associated swell**	
Significant wave height (m)	1.1	Significant wave height (m)	1.6
Direction (°)	45	Direction (°)	30
Period (sec)	6.3	Period (sec)	13.7
γ (JONSWAP)	1.8	σ (Gaussian spectral width)	0.007
Tide	1.2		
Wind		**Wind**	
Wind with wave, 10-min, 33' (kt)	14	Wind with wave, 10-min, 33' (kt)	27
Direction (°)	225	Direction (°)	225
Current (in line with wave)		**Current (in line with wave)**	
Surface speed (knot)	0.3	Surface speed (knot)	0.5
3 ft off bottom (knot)	0.3	3 ft off bottom (knot)	0.3

For a pipe-laying barge, the equation of motion is given by the following relationship:

$$[M]\{\ddot{\xi}\} + [C]\{\dot{\xi}\} + [K]\{\xi\} = \left\{F_{static} + F_{wavefreq} + F_{slowdrift} + F_{mooring}\right\} \quad (5.1)$$

where $[M]$ is the mass matrix, $[K]$ is the stiffness, $[C]$ is the damping matrix, and $\{\xi\}$ is the displacement vector. F_{static} is the force that arises from wave and wind; $F_{wavefreq}$ arises due to first-order wave effects; $F_{slowdrift}$ arises due to waves, the turbulence of the wind, and currents, which are acting at a frequency other than the usual wave frequency; and $F_{mooring}$ arises from the mooring lines. Xiong et al. (2015) proposed the governing equation for a mooring line as follows:

$$\frac{\partial^2 z}{\partial x^2} = \frac{w}{T_h}\left(\sqrt{1 + \left(\frac{\partial z}{\partial x}\right)^2}\right) \quad (5.2)$$

where w is the submerged weight per unit length of the line and T_h is the horizontal tension of the mooring line at (x, z). By ignoring the effects of elasticity, bending, and torsion stiffness, the following expression is valid:

$$z = \frac{T_h}{w}\left(\cosh\left(\frac{wx}{T_h} \right) - 1 \right) \tag{5.3}$$

By including the influence of elasticity in the problem, horizontal and vertical displacements are modified as follows:

$$dx = ds\left(1 + \frac{T}{AE} \right) \cos\varphi \tag{5.4}$$

$$dz = ds\left(1 + \frac{T}{AE} \right) \sin\varphi \tag{5.5}$$

where s is the length of the line segment at any point (x, z), T is the total tension along the line, and φ is the angle made by the tangent of the line to the horizontal at any point (x, z). At the touchdown point and beyond (until the anchor point), angle φ is zero. The tension component of the mooring line along the horizontal and vertical are given by:

$$T_h = AE\sqrt{\left(\left(\frac{T}{AE} + 1 \right)^2 - \frac{2wT_h}{AE} \right)} - AE \tag{5.6}$$

$$T_v = wL \tag{5.7}$$

The horizontal scope of the mooring line, from the fairlead to the touchdown point is given by:

$$X = \frac{T_h}{w}\sinh^{-1}\left(\frac{wL}{T_h} \right) + \left(\frac{T_h}{AE} \right) \tag{5.8}$$

The barge is operating at a shallow depth of 3 to 6 m, which is moored with an 8-point spread mooring system. Eight mooring lines are arranged around the vessel to form a spread mooring system with the fairlead points located at the corners of the vessel; two mooring lines originate from each corner. The mooring line arrangement is shown in Figure 5.17, while the mooring line properties are given in Table 5.4.

5.9 MOORING LINE ARRANGEMENT

In the present study, a pipe-laying barge is assumed to be operational in a shallow water depth of 15 m. A nominal diameter of 38 mm is used for the mooring line. The catenary equation is used to find the initial length of the mooring line. While the grade of the mooring line used is R4, CBS and proof loads for the line are obtained as per code guidelines. Mechanical properties of different steel grades

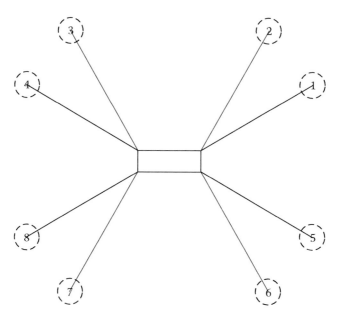

FIGURE 5.17   Mooring line layout.

TABLE 5.4
Mooring line properties

Description	Details
No. of mooring lines	8
Grade of steel	R4
Submerged weight of the line (N/m)	270.75
Axial stiffness (MN/m)	130
Length of mooring line (m)	650.0
Distance between center of barge and anchor (m)	656.6

are given in Table 5.5, while Table 5.6 summarizes the C value for different grades of steel as per International codes. The total length of the line is fixed after successfully determining the safety length. A proper decision on the safety length is crucial to prevent anchor uplift. Based on the results of tension variations in the existing mooring configuration, a new mooring layout will be proposed and assessed.

Numerical studies are carried out to determine the response of the barge under moored conditions while examining tension variation in the mooring line. Fine mesh is attempted in the simulation, while the maximum size of the element used in the mesh is 1.3 m.

TABLE 5.5
Mechanical properties of steel grade for mooring line

Grade	Yield stress (MPa)	Tensile strength (MPa)	Elongation (%)
R3	410	690	17
R3S	490	770	15
R4	580	860	12
R4S	700	960	12
R5	760	1000	12

TABLE 5.6
Mooring line C-value for different grades of steel

Steel grade	C-value	
	CBS	**Proof load**
3	22.3	14.8
3S	24.9	18.0
4	27.4	21.6
5	25.1	31.9

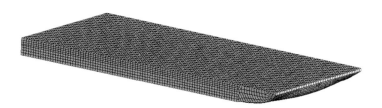

FIGURE 5.18 Numerical model of the barge.

The vessel is modeled using finite elements of 12446 in number with total nodes of 12439. The number of diffracting nodes and elements are about 6260 and 6114, respectively. The numerical model, simulated in the software, is shown in Figure 5.18.

5.10 WAVE-HEADING

The response time history of the pipe-laying barge in moored condition, under the influence of waves, is simulated in the time domain. The second-order differential equation of motion for the floating body is solved using a two-stage, semi-implicit predictor-corrector integration method to obtain the acceleration of the barge. The displaced position and velocity of the barge is obtained by integrating the accelerations at each time step. The moored barge is subjected to both drift and wave forces, which induce oscillations to the barge. The barge is subjected to drift motion, which is followed by wave motion. Response Amplitude Operator (RAO) for all degrees-of-freedom is computed for different heading angles. The solution obtained from the diffraction analyses is used as input for the hydrodynamic time-history response analyses. Wind and current effects are not considered in the analyses. Simulation is carried out for 2000 s at a time step of 0.05 s. RAO plots are generated for different wave-heading angles of 0°, 45°, and 90°. The influence of the low- and high-frequency motion of the barge in the pitch and surge motion and the tension variation in the mooring line are focused. In addition, the probability of the barge hitting the seafloor in shallow water due to massive heave displacement is also examined. As the barge is encountered by a wave heading angle of 0°, responses are seen in the surge, pitch, and heave degrees-of-freedom. Figure 5.19 shows the responses in all degrees-of-freedom; the figure shows that the sway, roll, and yaw responses are insignificant. Further, among all degrees-of-freedom, surge motion shows the maximum amplitude.

Figure 5.20 shows the response of the barge in all degrees-of-freedom, under a 45° wave-heading. Barge experiences wave load from all directions under a 45° wave heading angle; the response is seen in all degrees-of-freedom. The figure shows that among all response amplitudes, the surge and sway responses are dominant. Figure 5.21 shows the RAO of the barge in all active degrees-of-freedom under a 90° wave-heading. The response plots show the maximum amplitude in the sway motion. Heave motion is also activated, but surge, pitch, and yaw are insignificant.

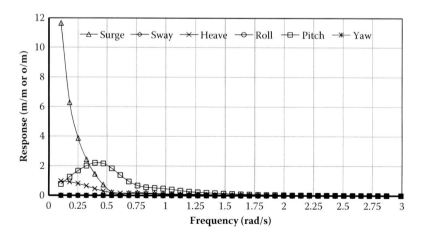

FIGURE 5.19   Barge response under 0° wave-heading.

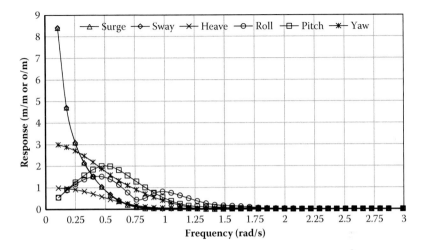

FIGURE 5.20   Barge response under 45° wave-heading.

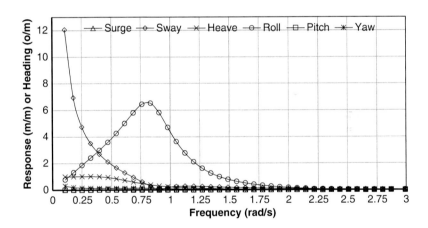

FIGURE 5.21   Barge response under 90° wave-heading.

It is observed from the above plots that the surge and sway responses have significant low-frequency content, arising from second-order drift forces. It confirms the fact that the catenary mooring gives sufficient compliancy to the barge, which is desirable for a smooth operation. While it reduces considerable tension variation in the mooring lines, it is still smaller in comparison to that of the taut-moored system. As tension will intervene with the compliancy of the barge, it is vital to examine the mooring line tension for its restoring capabilities.

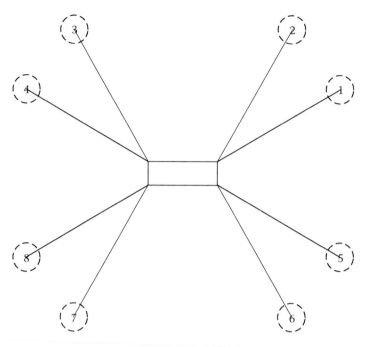

FIGURE 5.22   Mooring layout.

5.11 ANGLE BETWEEN MOORING LINES

The angles between the adjacent mooring lines are changed to 30°, 45°, and 60°, and the response of the barge is examined. Figure 5.22 shows the mooring line layout, while Figure 5.23 shows the tension variation in each one of the mooring lines for 30° between the adjacent mooring lines (for example, between mooring lines 1 and 2).

It is seen from the figures that the tension variation in each line is different even though the mooring line characteristics are similar. This is due to the arrangement of the mooring lines, proposed initially for the pipe-laying barge. It is observed that moorings lines 5 and 1 show the maximum amplitude under a 0° wave-heading; the corresponding maximum tension amplitudes are 1106.95 and 1162.36 kN, respectively. For the chosen mooring line configuration, the estimated breaking load is about 1600 kN (DNVOS-E301; E 302). It is seen that this value is not exceeded in any one of the mooring lines, under the given layout. Figure 5.24 shows the summary of tension amplitudes in the mooring lines.

Similar studies are also carried out for 45° and 60° arrangements. Figures 5.25 and 5.26 show the maximum tension in each mooring line for both the included angles. It is seen from the figures that mooring lines 1 and 5 show the maximum tension amplitude in both cases. For the 45° arrangement, maximum tension amplitudes are (1237, 1186), while for that of the 60°, it is about (1607, 1558), respectively. When compared with that of the 30° arrangement, the maximum

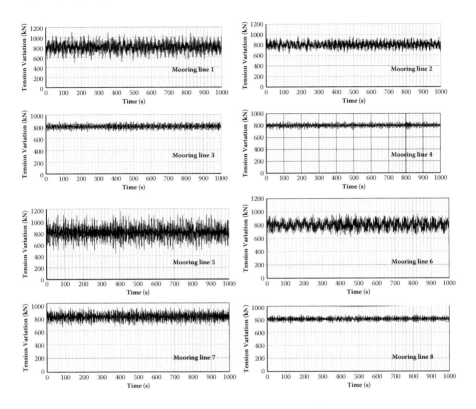

FIGURE 5.23   Tension variation in mooring lines (adjacent lines at 30° apart).

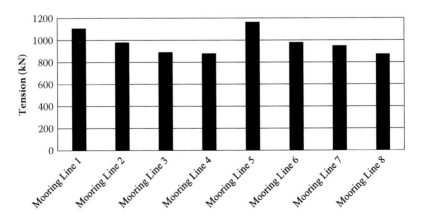

FIGURE 5.24   Maximum tension in each mooring line (30° apart).

amplitude of tension variation increases by about 6%, and 36% for 45° and 90°, respectively. It is because when the mooring lines are closer to each other, tension occurring due to the barge response is almost equal, and they are shared. On the

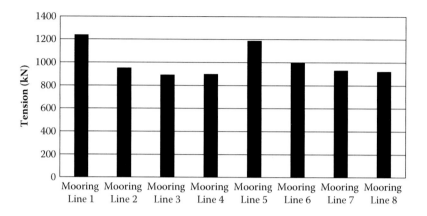

FIGURE 5.25   Maximum tension in each mooring line (45° apart).

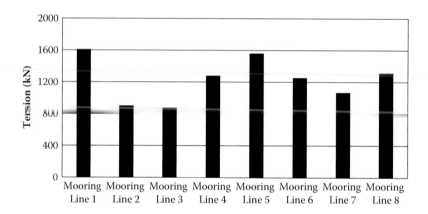

FIGURE 5.26   Maximum tension in each mooring line (60° apart).

contrary, when mooring lines are kept far apart, a single line has to cater to these tension variations. This increases the tension build-up in some particular mooring lines. Therefore, this imbalance in the tension variation needs to be investigated with a different layout.

5.12 STIFFNESS VARIATION

The influence of the stiffness variation on the response of the barge is also examined. Stiffness of the mooring lines is varied in the range of 130, 303, and 548 MN/m, respectively. Axial stiffness values are chosen for varying diameters of the mooring lines, which are feasible for a pipe-laying barge. For the head-sea condition, surge and pitch motion of the barge are analyzed and compared. Figure 5.27 shows the power spectral density plot (PSD) of the surge response under varying stiffness cases. It is

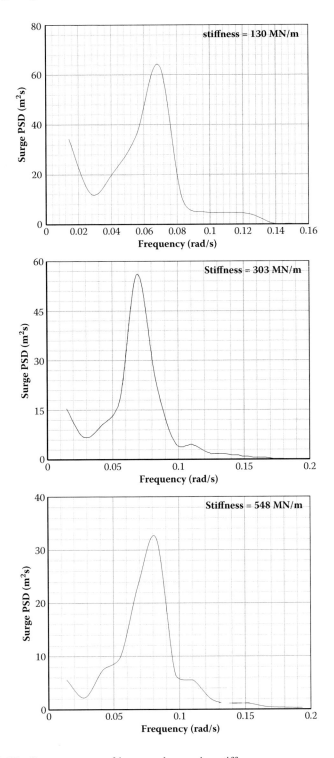

FIGURE 5.27   Surge response of barge under varying stiffnesses.

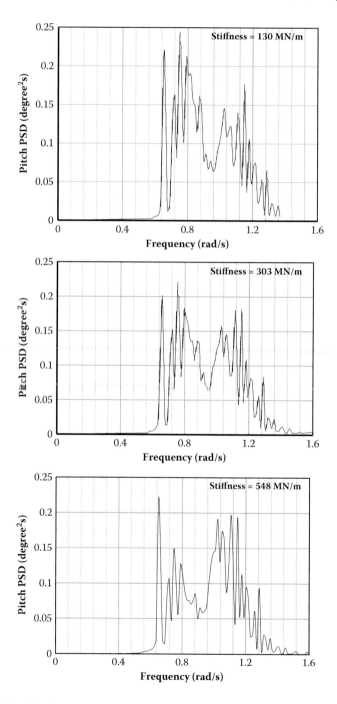

FIGURE 5.28 Pitch response of barge under varying stiffnesses.

TABLE 5.7

Comparison of barge responses under varying stiffnesses

Description	Surge	Sway	Heave	Roll	Pitch	Yaw
130 MN/m						
Max (m/^{0})	3.98	7.32	3.45	0.94	1.71	13.10
Min (m/^{0})	−1.74	−2.85	2.85	−1.03	−1.90	−3.05
303 MN/m						
Max (m/^{0})	3.23	6.51	3.44	0.93	1.67	11.37
Min (m/^{0})	−1.1847	−1.93	2.85	−0.91	−1.87	−1.52
% reduction	−16.37	−11.04	−0.40	−1.55	−2.02	−13.21
548 MN/m						
Max (m/^{0})	2.42	5.58	3.44	0.84	1.64	10.56
Min (m/^{0})	−1.04	−0.77	2.84	−0.84	−1.87	−0.37
% reduction	−44.89	−23.81	−0.56	-9.87	−3.79	−19.40

seen from the plots that variation in mooring line stiffness has a significant effect on the barge response in a horizontal plane; hence, only surge and pitch responses are focused. The variation in stiffness of mooring lines influences the restoring force; it increases with the increase in stiffness of the mooring line. It, in turn, restrains the vessel motion to the encountered waves. It is observed that the surge response of the barge decreases by about 16% with the increase in stiffness from 130 MN/m to 303 MN/m; a further decrease of about 45% is seen for an increase in stiffness to 548 MN/m. Figure 5.28 shows the PSD plots of the pitch response. It is seen from the plot that the change in stiffness in the mooring lines does not influence the pitch response of the barge, significantly. Table 5.7 shows a comparison of the barge response in various degrees-of-freedom under varying axial tension of the mooring lines; the percentage variation is also indicated.

5.13 WAVE PERIOD AND WAVE-HEADING

The wave period has a significant effect on the motion of the barge. While pipe-laying is carried out in shallow waters, a significant change is expected to occur in the heave motion; if not scrutinized, vessels may hit the seafloor. The threshold value for avoiding the bottom of the vessel touching the seabed is taken as a minimum of 1 m. Hence, the heave response of the barge, under moored condition, is examined under different wave periods and wave-heading angles. Figure 5.29 indicates the variations in heave response for different wave periods under different wave-heading angles. It is seen from the figure that the heave motion of the barge increases with an increase in the wave period. Among different wave-heading angles considered, 0° and 180° are the safest. As seen from the plots, at a 90° wave-heading angle, the threshold value is exceeded even for a minimal wave period of about 5 s.

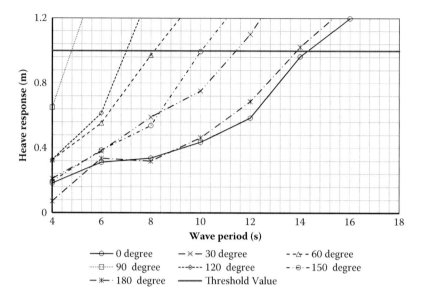

FIGURE 5.29   Heave response of barge under different wave-headings.

5.14 PROPOSED MOORING LAYOUT

Based on the analyses, as discussed above, a new mooring layout is proposed for the pipe-laying barge. Various parameters, namely, wave-heading angle, mooring line tension, mooring line stiffness, and wave period, are examined to propose the layout of mooring lines. Many factors influence the choice of designing an alternate layout of mooring lines for pipe-laying barges (API RP 2SK). They are as follows:

- A minimum vertical and horizontal clearance of 10 m is required at the crossing of mooring lines
- A minimum horizontal clearance of 10 m is required between the mooring line and other installations
- A minimum horizontal clearance of 100 m is required between the pipelines and mooring line anchor points.

Figure 5.30 shows an improved layout of the mooring line for the pipe-laying barge. The barge is analyzed under the new mooring layout for different wave-heading angles of 0° and 45°. Figure 5.31 shows the response of the barge under the improved mooring layout for the head-sea condition. It is seen from the figures that the maximum value of the response in surge motion is about 0.6 m. The heave response is about 0.17 m, and that of the pitch is about 3°. Figure 5.32 shows the response of the barge under a 45° wave-heading. The maximum response is 0.4 m for surge motion, 0.42 m for sway motion, 0.35 m for heave motion, 3.5° for roll motion, and 3° for pitch motion. It is interesting to note that responses of the barge in all active

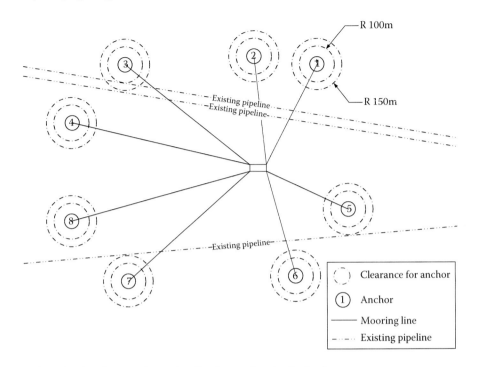

FIGURE 5.30 Improved mooring line layout for pipe-laying barge.

degrees-of-freedom, under the proposed layout of mooring lines, are significantly smaller than that of the original layout.

Tension variations in mooring lines are also examined under the improved layout of mooring lines. The maximum tensions in each mooring line, for both the wave-heading angles, are shown in Figure 5.33. It is seen from the figure that mooring line-2 experiences the maximum tension variation, which is about (1416, 1120 kN) for the 0° and 45° wave-headings, respectively. Both of these values are well below the minimum breaking load as prescribed by the standard codes for mooring lines. It is also important to note that tension variations under the improved layout of the mooring line are smaller than that of the original layout.

Table 5.8 shows the comparison of the barge response under the old and the new mooring layouts. It is seen from the table that the response of the barge, under the newly proposed layout of mooring lines, possesses a smaller response in comparison to that of the initial layout under similar sea states, which indicates better stability of the barge.

Figure 5.34 shows the comparison of two of the most critical mooring lines in the 45° wave-heading. It is clear from the chart that the newly suggested mooring layout possesses less dynamic tension compared to the initial layout when acted upon by the wave.

Under the standard layout of mooring lines, RAOs in the surge, pitch, and heave degrees dominate in 0° degree wave incidence, while that of sway, roll, and yaw

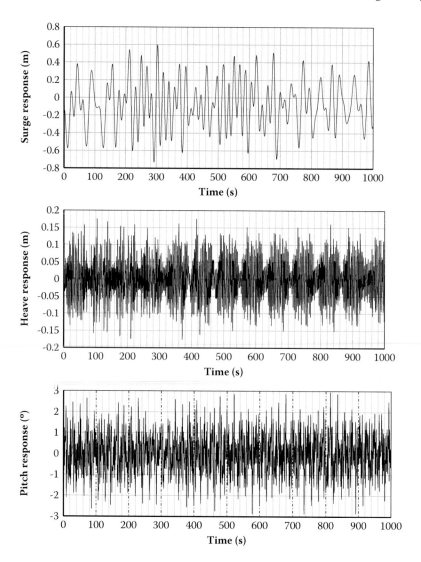

FIGURE 5.31   Barge response under improved mooring layout (0° wave-heading).

motion are insignificant. Surge response has significant low-frequency content, arising from second-order drift forces. Catenary mooring gives sufficient compliancy to the barge during operation. While it reduces considerable tension in mooring lines, it is still smaller in comparison to that of the taut moored system. The maximum value of tension variation increases by 6.37% and 36% for 45° and 90°, respectively, in comparison to that of the 0° wave-heading angle. It is because when mooring lines are closer to each other, they share tension caused by the vessel motion. Surge motion decreases by about 16% with an increase in stiffness from 130 MN/m to 303 MN/m; a further decrease of about 45% is seen for an increase in

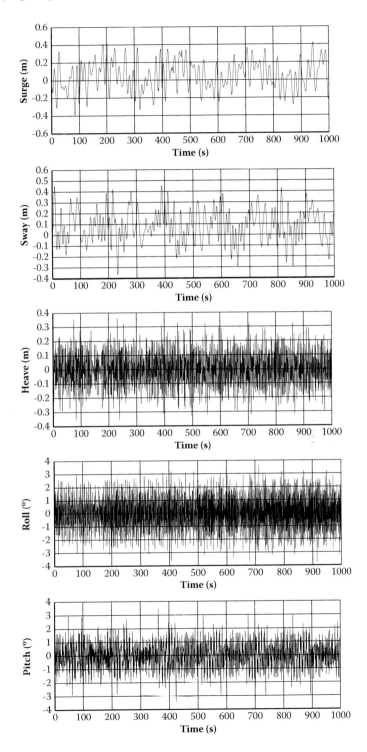

FIGURE 5.32 Barge response under improved mooring layout (45° wave-heading).

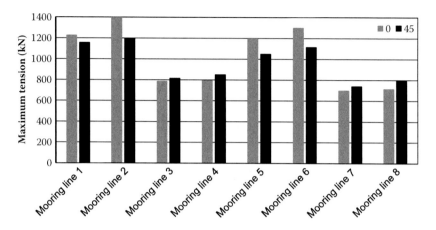

FIGURE 5.33 Maximum tension in each mooring line under the improved layout.

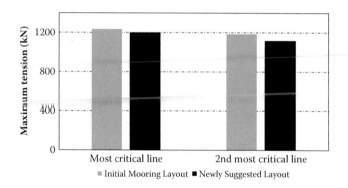

FIGURE 5.34 Critical lines in initial and newly suggested mooring layout.

TABLE 5.8
Comparison of barge responses under new and old layouts

	Surge	Sway	Heave	Roll	Pitch	Yaw
			Old layout			
Max (m/°)	3.48	7.26	3.48	0.92	1.74	12.36
Min (m/°)	−1.86	−2.24	2.82	-0.85	−1.82	−1.86
			New layout			
Max (m/°)	2.42	5.58	3.44	0.84	1.64	10.56
Min (m/°)	−1.04	−0.77	2.84	−0.84	−1.87	−0.37
% reduction	35.20599	33.15789	9.090909	5.084746	1.404494	23.13643

stiffness to 548 MN/m. Variation in the stiffness of mooring lines does not influence the pitch motion of the vessel significantly.

The heave motion of the barge increases with an increase in the wave period. Among wave-heading angles considered in the study, 90° shows the least safe condition while that of 0° and 180° are the safest. When the wave-heading angle is at 90°, the threshold value is exceeded even for a minimal wave period of about 5 s. Responses of the barge in all active degrees-of-freedom, under the improved layout of mooring lines, are significantly less than that of the original layout. Tension variations in mooring lines are also comparatively less. Further, maximum tension is less than the minimum breaking load as prescribed for mooring lines. An improved layout of mooring lines results in a reduced response of the pipe-laying barge.

References

Adrezin, R., Bar-Avi, P., and Benaroya, H. 1996. Dynamic response of compliant offshore structures — Review. *Journal of Aerospace Engineering*, 9: 114–131.

Al-Solihat, M.K. and Nahon, M. 2016. Stiffness of slack and taut moorings. *Ships and Offshore Structures*, 11(8): 890–904. DOI: 10.1080/17445302.2015.1089052.

American Petroleum Institute. 2007. *Interim Guidance on Hurricane Conditions in the Gulf of Mexico, USA*. API 2INT-MET, Washinton, DC.

API RP 2SK. 2018. *Design and Analysis of Station-Keeping Systems for Floating Structures*, 4th Ed. American Petroleum Institute, Washington, DC.

Bai, Y. and Bai, Q. (Eds.). 2005. *Subsea Pipelines and Risers*. Elsevier. ISBN: 0-080-4456-67.

Barltrop, N.D.P. 1998. *Floating Structures: A Guide for Design and Analysis*. CMPT and OPL, 2. ISBN: 1-870553-35-7.

Bjerager, P. 1990. On computation methods for structural reliability analysis. *Structural Safety*, 9(2): 79–96.

Bouws, E., Günther, H., Rosenthal, W., and Vincent, C. L. 1985. Similarity of the wind wave spectrum in finite depth water: 1. Spectral form. *Journal of Geophysical Research: Oceans*, 90: 975–986.

Cerik, B.C., Shin, H.K., and Cho, S.R. 2015. On the resistance of steel ring-stiffened cylinders subjected to low-velocity mass impact. *International Journal of Impact Engineering*, 84: 108–123.

Chandrasekaran, S. 2014. *Advanced Theory on Offshore Plant FEED Engineering*. Changwon National University Press, Republic of South Korea, 237. ISBN: 978-89-969-7928-9.

Chandrasekaran, S. 2015a. *Dynamic Analysis and Design of Ocean Structures*. Springer, India. ISBN: 978-81-322-2276-7.

Chandrasekaran, S. 2015b. *Advanced Marine Structures*. CRC Press, Florida (USA). ISBN 978-14-987-3968-9.

Chandrasekaran, S. 2016a. *Offshore Structural Engineering: Reliability and Risk Assessment*. CRC Press, Florida. ISBN: 978-14-987-6519-0.

Chandrasekaran, S. 2016b. *Health, Safety and Environmental Management for Offshore and Petroleum Engineers*. John Wiley & Sons, U.K. ISBN: 978-11-192-2184-5.

Chandrasekaran, S. 2017. *Dynamic Analysis and Design of Ocean Structures*, 2nd Ed. Springer, Singapore. ISBN:978-981-10-6088-5.

Chandrasekaran, S. 2018. *Advanced Structural Analysis with MATLAB*. CRC Press, Florida, USA. ISBN: 978-036-70-2645-5.

Chandrasekaran, S. 2019a. *Advanced Steel Design of Structures*. CRC Press, Florida, ISBN: 978-036-72-3290-0.

Chandrasekaran, S. 2019b. *Structural Health Monitoring with Application to Offshore Structures*. World Scientific Publishing Co., Singapore, ISBN: 978-971-12-0108-0.

Chandrasekaran, S. and Ajesh Kumar, P.T. 2017. Damage assessment in concrete marine structures using damage plasticity model. In *Proceedings of the 6th International Conference on Marine Structures (MARSTRUCT)*, 8–10 May, 2017, Lisbon, Portugal.

Chandrasekaran, S. and Ajesh Kumar, P.T. 2019. Damage detection in Reinforced concrete berthing jetty using a plasticity model approach. *Journal of Marine Science and Application*, 18: 482–491.

Chandrasekaran, S. and Bhattacharyya, S.K. 2012. *Analysis and Design of Offshore Structures with Illustrated Examples*. Human Resource Development Center for Offshore and Plant Engineering (HOPE Center), Changwon National University Press, Republic of Korea, p. 285. ISBN: 978-89-963-9155-5.

Chandrasekaran, S., Chandak, N.R., and Gupta, Anupam. 2006a. Stability analysis of TLP tethers. *Ocean Engineering*, 33(3): 471–482.

Chandrasekaran, S., Gaurav, Giorgio, Serino, and Salvatore, Miranda. 2011. Springing and ringing response of triangular TLPs. *International Shipbuilding Progress*, 58(2–3): 141–163.

Chandrasekaran, Gaurav, and Jain, A.K. 2009. Ringing response of offshore compliant structures. In *Proceedings of International Conference in Ocean Engineering (ICOE 2009)*, IIT Madras, India, 1–5th Feb 2009, pp. 55–56

Chandrasekaran, S., Gaurav, and Jain, A.K. 2010. Ringing response of offshore compliant structures. *International Journal of Ocean and Climate Systems*, 1(3–4): 133–144.

Chandrasekaran, S., Gaurav, and Srivastava, Shivam. 2008. Structural response of Offshore TLPs under seismic excitations. *International Engineering and Technology Journal of Civil and Structures*, 1(1): 7–12.

Chandrasekaran, S., Gaurav and Srivastava, Shivam. 2008a. Steady and transient response of triangular TLPs under random wave load. In *Seventh European Conference on Structural Dynamics (EuroDyn 2008)*, 7–9th July 2008, Southampton, U.K, pp. 50 (CD-ROM: Ref No. E64).

Chandrasekaran, S., Hari, S., and Amirthalingam, Murugaiyan. 2019. Wire-arc additive manufacturing of functionally graded material for marine riser applications. In *Proceedings of I-OCEANS 2019*, Universiti Teknologi Malaysia, Terengganu, Malaysia, 5–7th Aug 2019.

Chandrasekaran, S. and Jain, A.K. 2002a. Dynamic behavior of square and triangular offshore tension leg platforms under regular wave loads. *Ocean Engineering*, 29(3): 279–313.

Chandrasekaran, S. and Jain, A.K. 2002b. Triangular configuration tension leg platform behavior under random sea wave loads. *Ocean Engineering*, 29(15): 1895–1928.

Chandrasekaran, S. and Jain, A.K. 2016. *Ocean Structures: Construction, Materials and Operations*. CRC Press, Florida. ISBN: 978-149-81-9742-9.

Chandrasekaran, S., Jain, A.K., and Chandak, N.R. 2004. Influence of hydrodynamic coefficients in the response behavior of triangular TLPs in regular waves. *Ocean Engineering*, 31(17–18): 2319–2342.

Chandrasekaran, S., Jain, A.K., and Chandak, N.R. 2007a. Response behavior of triangular TLPs under regular waves using Stokes non-linear wave theory. *ASCE Journal of Waterway, Port, Coastal and Ocean Engineering*, 133(3): 230–237.

Chandrasekaran, S., Jain, A.K., and Chandak, N.R. 2006b. Seismic analysis of offshore triangular tension leg platforms. *International Journal of Structural Stability and Dynamics*, 6(1): 1–24.

Chandrasekaran, S., Jain, A.K., and Gupta, A. 2007b. Influence of wave approach angle on TLP's response. *Ocean Engineering*, 8–9(34): 1322–1327.

Chandrasekaran, S., Jain, A.K., Gupta, A., and Srivastava, A. 2007d. Response behavior of triangular tension leg platforms under impact loading. *Ocean Engineering*, 34(1): 45–53.

Chandrasekaran, S., Jain, A.K., and Madhuri, S. 2013a. Aerodynamic response of offshore triceratops. *Ship and Offshore Structures*, 8(2): 123–140. DOI: 10.1080/17445302.2012.691271.

Chandrasekaran, S. and Khader, Shihas A. 2016a. Hydrodynamic performance of a moored barge in regular waves. In *Proceedings of the 18th International Conference on Coastal and Ocean Engineering*, Jan 12–13, Zurich, Switzerland, pp. 687–694.

Chandrasekaran S. and Khader, Shihas A. 2016b. Hydrodynamic performance of a moored barge in irregular wave. *International Journal of Environmental, Chemical and Ecological Engineering*, 10(1): 47–54.

Chandrasekaran, S. and Kiran, P.A. 2017. Mathieu stability of offshore Buoyant Leg Storage and Regasification Platform. In *Proceedings of the 19th International Conference on*

Coastal and Ocean Engineering, World Academy of Science, Engg and Technology (WASET), Paris, France, 21–22 Sep. 2017, pp. 1605–1608.

Chandrasekaran, S. and Kiran, P.A. 2018a. Mathieu stability of buoyant leg storage and regasification platform. *Journal of Ocean Systems Engineering*, 8(3): 345–360. DOI: 10.12989/ose.2018.8.3.345.

Chandrasekaran S. and Kiran, P.A. 2018b. Mathieu stability of offshore triceratops under postulated failure. *Ships and Offshore Structures*, 13(2): 143–148. DOI: 10.1080/17445302.2017.133578.

Chandrasekaran, S. and Koshti, Yuvraj. 2013. Dynamic analysis of a tension leg platform under extreme waves. *Journal of Naval Architecture and Marine Engineering*, 10: 5968. DOI: 10.3329/jname.v10i1.14518.

Chandrasekaran, S., Kumar, Deepak, and Ramanathan, Ranjani. 2013c. Dynamic response of tension leg platform with tuned mass dampers. *Journal of Naval Architecture and Marine Engineering*, 10(2): 149–156.

Chandrasekaran, S. and Lognath, R.S. 2015. Dynamic analyses of buoyant leg storage regasification platform (BLSRP) under regular waves: Experimental investigations. *Ships and Offshore Structures*, 12(2): 227–232.

Chandrasekaran S. and Lognath, R.S. 2017. Dynamic analyses of buoyant leg storage and re-gasification platforms: Numerical studies. *Journal of Marine Systems and Ocean Technology*, 12(2): 39–48. DOI: 10.1007/s40868-017-0022-6.

Chandrasekaran, S., Lognath, R.S., and Jain, A.K. 2015c. Dynamic analysis of buoyant leg storage and regasification platform under regular waves. In *Proceedings of the 34th Conference on Ocean, Offshore and Arctic Engineering (OMAE 2015)*, St. John's, NL, Canada, May 31–June 5, 2015. OMAE2015-41554.

Chandrasekaran, S. and Madhavi, N. 2014a. Numerical study on geometrical configurations of perforated cylindrical structures. *Journal of Performance of Constructional Facilities, ASCE*. DOI: 10.1061/(ASCE)CF.1943-5509.0000687.

Chandrasekaran, S. and Madhavi, N. 2014b. Retrofitting of offshore structural member using perforated cylinders. *SFA Newsletter*, 13: 10–11.

Chandrasekaran, S. and Madhavi, N. 2015a. Design aids for offshore structures with perforated members. *Ship and Offshore Structures*, 10(2): 183–203. DOI: 10.1080/17445302.2014.918309.

Chandrasekaran, S. and Madhavi, N. 2015b. Retrofitting of offshore cylindrical structures with different geometrical configuration of perforated outer cover. *International Journal of Shipbuilding Progress*, 62(1–2): 43–56. DOI: 3233/ISP-150115.

Chandrasekaran, S. and Madhavi, N. 2015c. Flow field around an outer perforated circular cylinder under regular waves: Numerical study. *International Journal of Marine System and Ocean Technology*. DOI: 10.1007/s40868-015-0008-1.

Chandrasekaran, S. and Madhavi, N. 2015d. Estimation of force reduction on ocean structures with perforated members. In *Proceedings of the 34th International Conference on Ocean, Offshore and Arctic Engineering (OMAE2015)*, St. John's, NL, Canada, May 31–June 5, OMAE2015 – 41153.

Chandrasekaran, S. and Madhavi, N. 2016. Variation of flow field around twin cylinders with and without outer perforated cylinder: Numerical studies. *China Ocean Engineering*, 30(5): 763–771.

Chandrasekaran, S. and Madhavi, N. 2014c. Hydrodynamic performance of retrofitted structural member under regular waves. *International Journal of Forensic Engineering, Inder Science*, 2(2): 100–121.

Chandrasekaran, S. and Madhavi, N. 2014d. Variations of water particle kinematics of offshore TLP's with perforated members: numerical investigations. In *Proceedings of Structural engineering Convention 2014*, IIT Delhi, India. Dec 22–24, 2014.

Chandrasekaran, S. and Madhavi, N. 2014e. Variation of water particle kinematics with perforated cylinder under regular waves. In *Proceedings of ISOPE 2014*, 15–20 June, Busan, South Korea (paper accepted- 2014-TPC-0254).

Chandrasekaran, S., Madhavi, N., and Natarajan, C. 2014c. Variations of hydrodynamic characteristics with the perforated cylinder. In *Proceedings of the 33rd Conference on Ocean, Offshore and Arctic Engineering, OMAE 2014*, 8–13, June, San Francisco, USA (paper accepted: 23455).

Chandrasekaran, S., Madhavi, Natarajan and Saravanakumar. 2013b. Hydrodynamic response of offshore tension leg platforms with perforated members. *International Journal of Ocean and Climate Systems*, 4(3): 182–196.

Chandrasekaran, S. and Madhuri, S. 2012a. Free vibration response of offshore triceratops: Experimental and analytical investigations. In *3rd Asian Conference on Mechanics of Functional Materials and Structures (ACFMS)*, 8–9 Dec, IIT Delhi, pp. 965–968.

Chandrasekaran, S. and Madhuri, S. 2012b. Stability studies on offshore triceratops. In *Tech Samudhra International Conference on Technology of the Sea, Indian Maritime University*, Vishakapatnam, 1(10): 398–404.

Chandrasekaran, S. and Madhuri, S. 2015. Dynamic response of offshore triceratops: Numerical and Experimental investigations. *Ocean Engineering*, 109(15): 401–409. DOI: 10.1016/j.oceaneng.2015.09.042.

Chandrasekaran, S. and Madhuri, S. 2012c. Stability studies on offshore triceratops. *International Journal of Research and Development*, 1(10): 398–404.

Chandrasekaran, S., Madhuri, S., Jain, A.K., and Gaurav. 2010a. Dynamic response of offshore triceratops under environmental loads. In *Proceedings of International Conference of Marine Technology (MARTEC-2010)*, 11–12 Dec 2010 Dhaka Bangladesh, pp. 61–66.

Chandrasekaran, S. and Nagavinotihini, R. 2017a. Analysis and design of offshore triceratops under ultra-deep waters. *International Journal of Structural and Construction Engineering, World Academy of Science, Engineering and Technology*, 11(11): 1505–1513.

Chandrasekaran, S. and Nagavinothini, R. 2017b. Analysis and Design of offshore triceratops in ultra-deep waters. In *Proceedings of the International Conference of Offshore structures analysis and Design*, Nov. 29–30, Melbourne, Australia.

Chandrasekaran, S. and Nagavinothini, R. 2018a. Tether analyses of offshore triceratops under wind, wave and current. *Journal of Marine Systems and Ocean Technology*, 13: 34–42. DOI: 10.1007/s40868-018-0043-9.

Chandrasekaran, S. and Nagavinothini, R. 2018b. Dynamic analyses and preliminary design of offshore triceratops in ultra-deep waters. *International Journal of Innovative Infrastructure Solutions*, 3(1): 16. DOI: 10.1007/s41062-017-0124-1.

Chandrasekaran, S. and Nagavinothini, R. 2018c. Dynamic analysis of offshore triceratops under forces due to ice crushing in ultra-deep waters. In *Proceedings of the 11th International Conference on Marine Technology, MARTEC-2018*, Aug. 13–14, Kuala Lumpur, Malaysia.

Chandrasekaran, S. and Nagavinothini, R. 2019a. Response of triceratops to impact forces: numerical investigations. *Ocean Systems Engineering Journal, Techno Press*, 9(4): 349–368. DOI: 10.12989/ose.2019.9.4.349

Chandrasekaran, S. and Nagavinothini, R. 2019b. Ice-induced response of offshore triceratops. *Ocean Engineering*, 180: 71–96. DOI: 10.1016/j.oceaneng.2019.03.063.

Chandrasekaran, S. and Nagavinothini, R. 2019c. Tether analyses of offshore triceratops under ice loads due to continuous crushing. *International Journal of Innovative Infrastructure Solutions*, 4: 25. DOI: 10.1007/s41062-019-0212-5.

Chandrasekaran, S. and Nagavinothini, R. 2020a. Behaviour of stiffened deck plates under hydrocarbon fire. *Journal of Marine Systems and Ocean Technology*, 15: 95–109. DOI: 10.1007/s40868-020-00077-1.

Chandrasekaran, S. and Nagavinothini, R. 2020b. *Offshore Compliant Platforms: Analysis, Design and Experimental Studies*. Wiley, U.K., ISBN: 978-1-119-66977-7.

Chandrasekaran, S. and Nagavinothini, R. 2020c. Offshore triceratops under impact forces in ultra-deep arctic waters. *International Journal of Steel structures*. DOI: 10.1007/s13296-019-00297-.

Chandrasekaran, S. and Nagavinothini, R. 2020d. Parametric studies on the impact response of offshore triceratops in ultra-deep waters. *Structure and Infrastructure Engineering*, 16(7): 1002–1018. DOI: 10.1080/15732479.2019.1680707.

Chandrasekaran, S. and Nannaware, Madhuri. 2013. Response analyses of offshore triceratops to seismic activities. *Ship and Offshore Structures*, 9(6): 633–642. DOI: http://dx.doi.org/10.1080/17445302.2013.843816.

Chandrasekaran, S. and Nassery, Jamshed. 2015. Springing and ringing response of offshore triceratops. In *Proceedings of the 34th Conference on Ocean, Offshore and Arctic Engineering (OMAE 2015)*, St. John's, NL, Canada, May 31–June 5, 2015. OMAE2015-41551

Chandrasekaran, S. and Nassery, Jamshed. 2017a. Nonlinear response of stiffened triceratops under impact and non-impact waves. *International Journal of Ocean Systems Engineering*, 7(3): 179–193. DOI: 10.12989/ose.2017.7.3.179.

Chandrasekaran S. and Nassery, J. 2017b. Ringing response of offshore triceratops. *Journal of Innovative Infrastructure Solutions, Springer*. DOI: 10.1007/s41062-017-0092-5.

Chandrasekaran, S., Ranjani, R., and Kumar, Deepak. 2015a. Response control of tension leg platform with passive damper: Experimental investigations. *Ships and Offshore Structures* 12(2): 171–181.

Chandrasekaran, S. and Rao, Manda Hari Venkata Ramachandra. 2019. Numerical analysis on triceratops restraining system: failure conditions of tethers. In *Proceedings of the International Conference on Coastal and Ocean Engineering*, Sep 19–20, 2019, Paris, France.

Chandrasekaran, S. and Rao, M.H.V.R. 2019. Numerical Analysis on triceratops restraining system: failure conditions of tethers. *International Journal of Environmental and Ecological Engineering, World Academy of Science, Engineering and Technology*, 13(9): 588–592.

Chandrasekaran and Roy, Anubhab. 2004. Computational wave theories for deep water compliant offshore structures. In *Proceedings of International Conference on Environmental Fluid Mechanics (ICIFEM), IIT*, Guwahati, March 2–3, pp. 138–145.

Chandrasekaran, S. and Roy, Anubhab. 2005. Phase plane study of Offshore Structures subjected to nonlinear hydrodynamic loading. In *Proceedings of International Convention of Structural Engineering (SEC 2005)*, IISc Bangalore, pp. 397.

Chandrasekaran, S., Senger, Mayank, Jain, Arvind, 2015b. Dynamic response behavior of stiffened triceratops under regular waves: Experimental investigations. In *Proceedings of the 34th Conference on Ocean, Offshore and Arctic Engineering (OMAE 2015)*, St. John's, NL, Canada, May 31–June 5, 2015. OMAE2015-41376.

Chandrasekaran, S. and Senger, Mayank. 2017. Dynamic analyses of stiffened triceratops under regular waves: Experimental investigations. *Ships and Offshore Structures*, 12(5): 697–705.

Chandrasekaran, S., Serino, G., Jain, A.K., Salvatore, Miranda, Gupta, Anupam, Gaurav, and Sharma, Abhishek. 2008b. Influence of varying inertia coefficient and wave directionality on TLP geometry. *8th ISOPE Asia/Pacific Offshore Mechanics Symposium* (ISOPE-PACOMS-2008), 10–13th Nov 2008. Bangkok, Thailand.

Chandrasekaran, S. and Sharma, Abhishek. 2010. Potential flow based numerical study for the response of floating offshore structures with perforated columns. *Ships and Offshore Structures*, 5(4): 327–336.

Chandrasekaran, S., Sharma, Abhishek, and Srivastava, Shivam. 2007c. Offshore triangular TLP behavior using dynamic Morison equation. *Structural Engineering*, 34(4): 291–296.

Chandrasekaran, S. and Srivastava, Gaurav. 2008. Offshore Triangular TLP earthquake motion analysis under distinctly high sea waves. *Ship and Offshore Structures*, 3(3): 173–184.

Chandrasekaran, S. and Srivastava, G. 2017. *Design Aids for Offshore Structures under Special Environmental Loads Including Fire Resistance*. Springer, Singapore. ISBN 978-981-10-7608-4.

Chandrasekaran S. and Thailammai, C.T. 2018. Structural health monitoring of offshore Buoyant leg storage and re-gasification platform: Experimental investigations. *Journal of Marine Science and Application*, 17: 87–100. DOI: 10.1001/s11804-018-0013-9.

Chandrasekaran, S., Thaillammai, C., and Khader, Shihas A. 2016a. Structural health monitoring of offshore structures using wireless sensor networking under operational and environmental variability. *International Journal of Environmental, Chemical and Ecological Engineering*, 10(1): 33–39.

Chandrasekaran, S. and Thomas, Merin. 2016. Suppression system for offshore cylinders under vortex induced vibration. In *Proceedings of the 21st Vibro Engineering Conference*, Berno, Czech Republic, 31st Aug–1st Sep, 7: 1–6.

Chandrasekaran, S. and Uddin, S.A. 2020. Postulated failure analyses of a spread-moored semi-submersible. *Innovative Infrastructure Solutions*, 5: 36. DOI: 10.1007/s41062-020-0284-2.

Chandrasekaran, S. and Vishruth, Srinath. 2013. Experimental investigations of dynamic response of tension leg platforms with perforated members. In *Proceedings of the 32nd International Conference on Ocean, Offshore and Arctic Engineering, OMAE 2013*, Nantes, France, 9–14th June, 2013, OMAE 2013-10607.

Chen, P., Chai, S., and Ma, J. 2011. Performance evaluations of taut-wire mooring systems for the deep-water semi-submersible platform. In *Proceedings of the 30th International Conference on Ocean, Offshore and Arctic Engineering*, Rotterdam, The Netherlands, pp. 207–215.

Hang S. Choi, Hyun S. Shin, Park I. K., Rho, and Jun B. 2003. An experimental study for mooring effects on the stability of the SPAR platforms. In *Proceedings of the Thirteenth International Offshore and Polar Engineering Conference*, Honolulu, Hawaii, 1: 285–288.

Cho, S.R., Choi, S.I., and Son, S.K. 2015. Dynamic material properties of marine steels under impact loadings. In *Proceedings of the 2015 World Congress on Advances in Structural Engineering and Mechanics*, Incheon, Korea, Aug. 25–29.

Clauss, G., Schmittner, C., and Stutz, K. 2002. Time-domain investigation of a semi-submersible in rogue waves. In *Proceedings of the 21st International Conference on Offshore Mechanics and Arctic Engineering*, Oslo, Norway, June 23–28, pp. 509–516.

Clauss, G.F., Vannahme, M., Ellerman, K., and Kreuzer, E. 2000. Subharmonic oscillations of moored floating cranes. In *Offshore Technology Conference*, Houston, Texas, May 1–4, pp. 1–8.

Davenport, A.G. 1961. The spectrum of horizontal gustiness near the ground in high winds. *Quarterly Journal of the Royal Meteorological Society*, 87(372): 194–211. DOI: 10.1002/qj.49708737208.

Dean, R.G. and Dalrymple, R.A. 1991. Water wave mechanics for scientists and engineers. *World Scientific, Advanced Series on Ocean Eng*, 2: 170–186. ISBN: 978-981-02-0421-1.

Det Norske Veritas. 2016a. *DNV — Recommended Practice C203: Fatigue design of offshore steel structures.* Group Communications, Oslo, Norway.

Det Norske Veritas. 2016b. *DNV — Recommended Practice F205: Global Performance Analysis of Deepwater Floating Structures.* Group Communications, Oslo, Norway.

Det Norske Veritas. 2008. *DNV — Recommended Practice DNV-OS-E301: Position mooring*, pp. 34–34. Group Communications, Oslo, Norway.

Dev, A.K. and Pinkster, J.A. 1995. Viscous mean drift forces on moored semi-submersibles. In *Fifth International Offshore and Polar Engineering. Conference, International Society of Offshore and Polar Engineers*, Hague, The Netherlands, June 11–16.

El-gamal, A.R., Essa, A., and Ismail, A. 2014. Tethers tension force effect in the response of a squared tension leg platform subjected to ocean waves. *Ocean Systems Engineering*, 4(4): 327–342.

Ellermann, K. 2005. Dynamics of a moored barge under periodic and randomly disturbed excitation. *Ocean Engineering*, 32: 1420–1430.

Ellermann, K. and Kreuzer, E. 2000. Moored crane vessels in regular waves. In *Proceedings of UTAM Symposium on Recent Developments in Non-linear Oscillations of Mechanical Systems*, Hanoi, Vietnam, March 2–5, pp. 105–113.

Ellermann, K., Kreuzer, E., and Markiewicz, M. 2002. Nonlinear dynamics of floating cranes. *Nonlinear Dynamics*, 27: 107–183.

Felix-Gonzalez, I. and Mercier, R.S. 2016. Optimized design of statically equivalent mooring systems. *Ocean Engineering*, 111: 384–397.

Gao, Y., Zong, Z., and Sun, L., 2011. Numerical prediction of fatigue damage in steel catenary riser due to vortex-induced vibration. *Journal of Hydrodynamics*, 23(2): 154–163.

Garza-Rios, L.O. and Bernitsas, M.M. 1996. Analytical expressions of the stability and bifurcation boundaries for general spread mooring systems. *Journal of Ship Research*, 40: 337–350.

Garza-Rios, L.O., Bernitsas, M.M., Nishimoto, K., and Matsuura, J.O.P.J. 2000. Dynamics of spread mooring systems with hybrid mooring lines. *Journal of Offshore Mechanics and Arctic Engineering*, 122(4): 274–281.

Gottlieb, O. and Yim, S.C. 1992. Nonlinear oscillations, bifurcations and chaos in a multi-point mooring system with a geometric nonlinearity. *Applied Ocean Research*, 14(4): 241–257.

Gouveia, J., Sriskandarajah, T., Karunakaran, D., Manso, D., Chiodo, M., Zhou, D., and Escudero, C. 2015. Steel Catenary Risers (SCRs): from Design to Installation of the First Reeled CRA Lined Pipes. Part I — Risers Design. In *Offshore Technology Conference*, Houston, TX, USA.

Guédé, F. 2017. Risk-based structural integrity management for offshore jacket platforms. *Marine Structures*, 63: 223–241.

Hallam M.G., Heaf N.J., and Wootton L.R., 1977. Dynamics of Marine Structures. *CIRIA Underwater Engineering Group, Report UR8 Atkins Research and Development*, British Ship Research Association, London, U.K.

Harding, J.E., Onoufriou, A., and Tsang, S.K. 1983. Collisions—What is the danger to offshore rigs. *Journal of Constructional Steel Research*, 3(2): 31–38.

Haritos, N. 2007. Introduction to the analysis and design of offshore structures–an overview. *Electronic Journal of Structural Engineering*, 7: 55–65.

Harris, R.I. 1971. The nature of the wind. The modern design of wind-sensitive structures. CRI, London, 29–55.

Hassan, A., Downie, M. J., Incecik, A., Baarholm, R., Berthelsen, P. A., Pakozdi, C., and Stansberg, C.T. 2009. Contribution of the mooring system to the low-frequency motions of a semisubmersible in combined wave and current. In *Proceedings of the 28th International Conference on Ocean, Offshore and Arctic Engineering*, pp. 55–62.

Hong, K.S. and Shah, U.H. 2018. Vortex-induced vibrations and control of marine risers: A review. *Ocean Engineering*, 152: 300–315.

Hover F.S., Tvedt H., and Triantafyllou M.S., 2001. Vortex induced vibrations of a cylinder with tripping wires. *Journal of Fluid Mechanics*, 448: 175–195.

How, B.V.E., Ge, S.S., and Choo, Y.S. 2009. Active control of flexible marine risers. *Journal of Sound and Vibration*, 320(4–5): 758–776.

Huang, T. and Dareing, D.W. 1968. Buckling and lateral vibration of drill pipe. *Journal of Engineering for Industry*, 90(4): 613–619. DOI: 10.1115/1.3604697.

Huang, W., Liu, H.X., Shan, G.M., and Hu, C., 2011. Fatigue analysis of the taut wire mooring system applied for deep waters. *Journal of China Ocean Engineering*, 25(3): 413–426. DOI: 10.1007/s13344-011-0034-5.

Hussain, A., Nah, E., Fu, R., and Gupta, A. 2009. Motion comparison between a conventional deep draft semi-submersible and a dry tree semi-submersible. In *Proceedings of the 28th International Conference on Ocean, Offshore and Arctic Engineering*, pp. 785–792.

Islam, A.B., Soeb, M.R., and Jumaat, M.Z. 2017. Floating SPAR platform as an ultra-deepwater structure in oil and gas exploration. *Ships and Offshore Structures*, 12(7): 923–936.

Jain, A.K. and Chandrasekaran, S. 2004. Aerodynamic behavior of Offshore triangular TLPs. *In Proceedings of 14th International Society of Offshore and Polar Engineering conference (ISOPE)*, Toulon, France, May 23–28, pp. 564–569.

Jain, A.K. and Chandrasekaran, S. 2004. Aerodynamic behaviour of Offshore triangular TLPs. In *Proceedings of 14th International Society of Offshore and Polar Engineering conference (ISOPE)*, Toulon, France, May 23–28, pp. 564–569.

Jang, B.S., Kim, J.D., Park, T.Y., and Jeon, S.B. 2019. FEA based optimization of semi-submersible floater considering buckling and yield strength. *International Journal of Naval Architecture and Ocean Engineering*, 11(1): 82–96.

Jeans, G., Grant, C., and Feld, G. 2002. Improved current profile criteria for deepwater riser design. In Proceedings of the 21st International Conference on Offshore Mechanics and Arctic Engineering, pp. 307–311.

Jensen, G.A., Säfström, N., Nguyen, T.D., and Fossen, T.I. 2010. A nonlinear PDE formulation for offshore vessel pipeline installation. *Ocean Engineering*, 37(4): 365–377.

Kahla, N.B. 1994. Dynamic analysis of guyed towers. *Engineering Structures*, 16(4): 293–301.

Kaimal, J.C., Wyngaard, J.C.J., Izumi, Y., and Coté, O.R. 1972. Spectral characteristics of surface-layer turbulence. *Quarterly Journal of the Royal Meteorological Society*, 98(417): 563–589.

Karmazinova, M.A., and Melcher, J.I. 2012. Influence of Steel Yield Strength Value on Structural Reliability. *Recent Researches in Environmental and Geological Sciences*. WSEAS Press, Greece, pp. 441–446.

Karna, T., Qu, Y., Bi, X., Yue, Q., and Kuehnlein, W. (2007). A spectral model for forces due to ice crushing. *Journal of Offshore Mechanics and Arctic Engineering*, 129(2): 138–145. DOI: 10.1115/1.2426997.

Karperaki, A.E., Belibassakis, K.A., and Papathanasiou, T.K. 2016. Time-domain, shallow-water hydroelastic analysis of VLFS elastically connected to the seabed. *Marine Structures*, 48: 33–51.

Kashiwagi, M. 2000. A time-domain mode-expansion method for calculating transient elastic responses of a pontoon-type VLFS. *Journal of Marine Science and Technology*, 5: 89–100.

Khalak A., Williamson C.H.K., 1991. Motions, forces and mode transitions in vortex induced vibrations at low mass-damping. *Journal of Fluids and Structures*, 13: 813–851.

Khan, N.U., and Ansari, K.A. 1986. On the dynamics of a multicomponent mooring line. *Computers and Structures*, 22: 311–334.

Kim, C.S., and Hong, K.S. 2009. Boundary control of container cranes from the perspective of controlling an axially moving string system. *International Journal of Control, Automation and Systems*, 7(3): 437–445.

Kim, Y., Kim, M.S., and Park, M.J. 2019. Fatigue analysis on the mooring chain of a spread moored FPSO considering the OPB and IPB. *International Journal of Naval Architecture and Ocean Engineering*, 11(1): 178–201.

King R., 1977. A review of vortex shedding research and its application. *Journal of Ocean Engineering*, 4: 141–171.

Konovessis, D., Chua, K.H., and Vassalos, D. 2014. Stability of floating offshore structures. *Ships and Offshore Structure*, 9(2): 125–133.

Kurian, V.J., Yassir, M.A., and Harahap, I.S., 2010. Nonlinear coupled dynamic response of a Semi-submersible platform. In *Proceedings of the 20th International Offshore and Polar Engineering Conference*, Beijing, China, June 20–25.

Kvitrud, A. 2011. Collisions between platforms and ships in Norway in the period 2001–2010. In *Proceedings of the 30th International Conference on Ocean, Offshore and Arctic Engineering, Rotterdam*, the Netherlands, June 19–24, pp. 637–641.

Lee, Y., Incecik, A., and Chan, H.S. 2005. Prediction of global loads and structural response analysis on a multi-purpose semi-submersible. In *Proceedings of the 24th International Conference on Offshore Mechanics and Arctic Engineering*, pp. 3–13.

Leira, B.J., Meling, T.S., Larsen, C.M., Berntsen, V., Stahl, B., and Trim, A. 2005. Assessment of fatigue safety factors for deep-water risers in relation to VIV. *Journal of Offshore Mechanics and Arctic Engineering*, 127(4): 353–358.

Lesage F., and Garthshore L.S., 1987. A Method of Reducing Drag and Fluctuating Side Force on Bluff Bodies. *Journal of Wind Engineering and Industrial Aerodynamics*, 25: 229–245.

Li, L. 2008. Floating structure responses in shallow water rough sea. In *Proceedings of the 18th International Offshore and Polar Engineering Conference*, Vancouver, BC, Canada, July 6–11, pp. 212–222.

Li, X., Yang, J., and Xiao, L. 2003. Motion analysis on a large FPSO in shallow water. In *Proceedings of the 13th International Offshore and Polar Engineering Conference*, Honolulu, Hawaii, USA, May 25–30, pp. 235–239.

Liapis S, Bhat S, Caracostis C, Webb C, and Lohr C. 2010. Global performance of the Perdido SPAR in waves, wind and current: Numerical predictions and comparison with experiments. In *Proceedings of the 29th International Conference on Ocean, Offshore and Arctic Engineering*, pp. 863–873.

Lie, H., and Kaasen, K.E. 2006. Modal analysis of measurements from a large-scale VIV model test of a riser in linearly sheared flow. *Journal of Fluids and Structures*, 22(4): 557–575.

Manco, M.R., Vaz, M.A., Cyrino, J.C., and Landesmann, A. (2013). Behavior of stiffened panels exposed to fire. In *Proceedings of IV MARSTRUCT*, Espoo, Finland, 101–108. ISBN 978-1-138-00045-2.

Marsh, G., Wignall, C., Thies, P.R., Barltrop, N., Incecik, A., Venugopal, V., and Johanning, L. 2016. Review and application of Rainflow residue processing techniques for accurate fatigue damage estimation. *International Journal of Fatigue*, 82: 757–765.

Mavrakos, S.A., and Chatjigeorgiou, J. 1997. Dynamic behavior of deepwater mooring lines with submerged buoys. *Computers and Structures*, 64(1–4): 819–835. DOI: 10.1016/S0045-7949(96)00169-1.

Mavrakos, S.A., Papazoglou, V.J., Triantafyllou, M.S., and Hatjigeorgiou, J. 1996. Deepwater mooring dynamics. *Marine Structures*, 9(2): 181–209. DOI: 10.1016/0951-8339(94)00019-O.

Mazzilli, C.E., and Sanches, C.T. 2011. Active control of vortex-induced vibrations in offshore catenary risers: a nonlinear normal mode approach. *Journal of Mechanics of Materials and Structures*, 6(7): 1079–1088.

McCoy, T.J., Brown, T., and Byrne, A. (2014). Ice Load Project Final Technical Report (No. DDRP0133). DNV GL, Seattle, WA (United States). DOI: 10.2172/1303304.

Melchers, R.E. 1988. Importance sampling in structural systems. *Structural Safety*, 6(1): 03–10.

Modarres-Sadeghi, Y., Chasparis, F., Triantafyllou, M.S., Tognarelli, M., and Beynet, P. 2011. Chaotic response is a generic feature of vortex-induced vibrations of flexible risers. *Journal of Sound and Vibration*, 330(11): 2565–2579.

Moo- Hyun Kim., 2013. *SPAR Platforms: Technology and Analysis Methods*. American Society of Civil Engineers. Washington, DC, ISBN-13:978-0784412091.

Nagavinothini R. and Chandrasekaran, S. 2019a. Dynamic analyses of offshore triceratops in ultra-deep waters under wind, wave and current. *Structures*, 20: 279–289.

Nagavinothini R, Chandrasekaran S. 2019b. Dynamic analyses of offshore triceratops in ultra-deep waters under the wind, wave, and current. *Structures*, 20: 279–289. Elsevier. DOI: 10.1016/j.istruc.2019.04.009.

Nagavinothini, R. and Chandrasekaran, S. 2020. Dynamic response of offshore Triceratops with elliptical buoyant legs. *International Journal of Innovative Infrastructure Solutions*, 5(47): 1–14. DOI: 10.1007/s41062-020-00298-8.

Ngo, Q.H., and Hong, K.S. 2010. Sliding-mode anti-sway control of an offshore container crane. *IEEE/ASME Transactions on Mechatronics*, 17(2): 201–209.

Ngo, Q H , and Hong, K S. 2012. Adaptive sliding mode control of container cranes. *IET Control Theory and Applications*, 6(5): 662–668.

Nguyen, D.H., Nguyen, D.T., Quek, S.T., and Sørensen, A.J. 2010. Control of marine riser end angles by position mooring. *Control Engineering Practice*, 18(9): 1013–1021.

Niedzwecki, J.M., and Liagre, P.Y. 2003. System identification of distributed-parameter marine riser models. *Ocean Engineering*, 30(11): 1387–1415.

Odijie, A.C., Wang, F., and Ye, J. 2017. A review of floating semisubmersible hull systems: Column stabilized unit. *Ocean Engineering*, 144: 191–202.

Onoufriou, T. 1999. Reliability-based inspection planning of offshore structures. *Marine Structures*, 12(7–8): 521–539.

Oppenheim, B. and Fletter, J. 1991. Design notes on spread mooring systems. In *Proceedings of the 1st International Offshore and Polar Engineering Conference*, Edinburg, The United Kingdom, Aug. 11–16, 2: 259–265.

Ormberg, H., and Larsen, K. 1998. Coupled analysis of floater motion and mooring dynamics for a turret-moored ship. *Applied Ocean Research*, 20(1–2): 55–67. DOI: 10.1016/S0141-1187(98)00012-1.

Pedersen, P.T. 1975. Equilibrium of offshore cables and pipelines during laying. *International Shipbuilding Progress*, 22: 399–408.

Phifer, E.H., Kopp, F., Swanson, R.C., Allen, D.W., and Langner, C.G. 1994. Design and installation of auger steel catenary risers. In *Proceedings of Offshore Technology Conference*, Houston, Texas, 3–6th May. DOI: 10.4043/7620-MS.

Prucz, Z. and Soong, T.T. 1984. Reliability and safety of tension leg platforms. *Engineering Structures*, 6(2): 142–149.

Qiao, D. and Ou, J. 2013. Global responses analysis of a semi-submersible platform with different mooring models in the South China Sea. *Ships and Offshore Structures*, 8(5): 441–456. DOI: 10.1080/17445302.2012.718971.

Qiao, D., Wu, F, and Ou, J., 2014. Coupled Analysis of a Semisubmersible platform with two types of mooring systems. In *Proceedings of the 24th International Society of Ocean and Polar Engineering Conference*, Busan, South Korea, June, 15–20.

Ranganathan, R. 1999. Structural reliability analysis and design. Jaico Publishing House, Mumbai, India, ISBN: 978-8172248512.

Reddy, D.V., and Swamidas, A.S. 2016. *Essentials of offshore structures: framed and gravity platforms*. CRC Press, Florida, ISBN: 9781420068825.

Rivera, M.R.M., Vaz, M.A., Cyrino, J.C.R., and Landesmann, A. 2014. Analysis of Oil Tanker Deck Under Hydrocarbon Fire. *International Journal of Modeling and Simulation for the Petroleum Industry*, 8(2).

Roberts, J.B. 1981. Non-linear analysis of slow drift-oscillations of moored vessels in random seas. *Journal of Ship Research*, 25: 130–140.

Rustad, A.M., Larsen, C.M., and Sørensen, A.J. 2008. FEM modelling and automatic control for collision prevention of top tensioned risers. *Marine Structures*, 21(1): 80–112.

Sadowski, A.J., Rotter, J.M., Reinke, T., and Ummenhofer, T. 2015. Statistical analysis of the material properties of selected structural carbon steels. *Structural Safety*, 53: 26–35.

Saha, G.K., Suzuki, K., and Kai, H. 2004. Hydrodynamic optimization of ship hull forms in shallow water. *Journal of Marine Science and Technology*, 9: 51–62.

Salama, M.M., Johnson, D.B., and Long, J.R. 1998. Composite production riser testing and qualification. *SPE Production and Facilities*, 13(03): 170–177.

Salama, M.M., Stjern, G., Storhaug, T., Spencer, B., and Echtermeyer, A. 2002. The first offshore field installation for a composite riser joint. In *Proceedings of Offshore Technology Conference*, Houstoc, TX, USA, 6–9th May. DOI: 10.4043/14018-MS.

Shih, L.Y. 1991. Analysis of ice-induced vibrations on a flexible structure. *Applied Mathematical Modelling*, 15(11–12): 632–638. DOI: 10.1016/S0307-904X(09)81009-3.

Sigurdsson, H. 1988. Gas Bursts from Cameroon Crater Lakes: A New Natural Hazard. *Disasters, Blackwell Publishing Ltd*, 12: 131–146. DOI: 10.1111/j.1467-7717.1988.tb00661.x.

Silvestre N. 2008. Buckling behavior of elliptical cylindrical shells and tubes under compression. *International Journal of Solids and Structures*, 45(16): 4427–4447.

Simiu, E. and Scanlan, R.H. 1978. Wind effects on structures: an introduction to wind engineering. Wiley, New York. ISBN-13: 978-0471866138.

Singh, N.K., Cadoni, E., Singha, M.K., and Gupta, N.K. 2011. Dynamic tensile behavior of multi-phase, high yield strength steel. *Materials and Design*, 32(10): 5091–5098.

Skrzypczak, I., Sáowikb, M., and Buda-Ołóga, L. 2017. The application of reliability analysis in engineering practice–reinforced concrete foundation. *Procedia Engineering*, 193: 144–151.

Sodhi, D.S., and Haehnel, R.B. 2003. Crushing ice forces on structures. *Journal of Cold Regions Engineering*, 17(4): 153–170. DOI: 10.1061/(ASCE)0887-381X(2003)17:4(153).

Sparks, C.P. 2007. Fundamentals of marine riser mechanics: Basic principles and simplified analyses. PennWell Books. ISBN: 978-1-59370-070-6.

Srinivasan, N., Chakrabarti, S., and Radha, R. 2006. Response analysis of a truss-pontoon semisubmersible with heave-plates. *Journal of Offshore Mechanics and Arctic Engineering*, 128(2): 100–107. DOI: 10.1115/1.2185679.

Stansberg, C.T. 2008. Current effects on a moored floating platform in a sea state. In *27th International Conference on Offshore Mechanics and Arctic Engineering*, American Society of Mechanical Engineers Digital Collection, Estoril, Portugal, June 15–20, pp. 433–444.

Sunil, D. K. and Mukhopadhyay, M. 1995. Free vibration of semisubmersibles: A parametric study. *Ocean Engineering*, 22(5): 489–502.

Teigen, P. 2005. Motion response of a spread moored barge over a sloping bottom. In *Proceedings of 15th International Offshore and Polar Engineering Conference*, Seoul, Korea, June 19–24, pp. 396–405.

Tennyson R.C., Booton M., and Caswell R.D. 1971. Buckling of imperfect elliptical cylindrical shells under axial compression. *AIAA Journal*, 9(2): 250–255.

Tim Finnigan, T.D., Irani M.B., and Van Dijk, R.R.T. 2005. Truss SPAR VIM in waves and currents. In *Proceedings of the 24th International Conference on Offshore Mechanics and Arctic Engineering*, June 12–17, Halkidiki, Greece. DOI: 10.1115/OMAE2005-67054.

Tran, A.M.D., Yoon, J.I., and Kim, Y.B. 2016. Motion control system design for a barge type vessel moored by ropes. In *AETA 2015: Recent Advances in Electrical Engineering and Related Sciences*, 371: 389–397.

Van Oortmerssen, G. 2008. The motions of a moored ship in waves. *Social Research*, 510: 131–132.

Verhagen, J.H.G. 1970. Low-frequency drifting force on a floating body in waves. *International Shipbuilding Progress*, 17: 136–145.

Wang, K., Er, G.K., and Iu, V.P. 2018. Nonlinear vibrations of offshore floating structures moored by cables. *Ocean Engineering*, 156: 479–488.

Watanabe, E., Utsunomiya, T., and Tanigaki, S. 1998. A transient response analysis of a very large floating structure by finite element method. *Doboku Gakkai Ronbunshu*, 598: 1–9.

Webster, W.C., 1995. Mooring induced damping. *Ocean Engineering*, 22(6): 571–591.

Wu, Y., Wang, T., Eide, Ø., and Haverty, K., 2015. Governing factors and locations of fatigue damage on mooring lines of floating structures. *Ocean Engineering*, 96: 109–124. DOI: 10.1016/j.oceaneng.2014.12.036.

Xiao, L., Tao, L., Yang, J., and Li, X. 2014. An experimental investigation on wave run-up along the broadside of a single point moored FPSO exposed to oblique waves. *Ocean Engineering*, 88: 81–90.

Xiong, L., Lu, H., Yang, J., and Zhao, W. 2015. Motion responses of a moored barge in shallow water. *Ocean Engineering*, 97: 207–217.

Xu, S., Ji, C.Y., and Soares, C.G. 2018. Experimental and numerical investigation a semi-submersible moored by hybrid mooring systems. *Ocean Engineering*, 163:641–678.

Xue, X., Chen, N.Z., Wu, Y., Xiong, Y., and Guo, Y. 2018. Mooring system fatigue analysis for a semi-submersible. *Ocean Engineering*, 156: 550–563.

Yan, J., Qiao, D., and Ou, J., 2018. Optimal design and hydrodynamic response analysis of deep-water mooring system with submerged buoys. *Journal of Ships and Offshore Structures*, 13(5): 476–487. DOI: 10.1080/17445302.2018.1426282.

Yan, J., Qiao, D., Fan, T., and Ou, J. 2016. Concept design of deepwater catenary mooring lines with submerged buoy. In *The 12th ISOPE Pacific/Asia Offshore Mechanics Symposium*. International Society of Offshore and Polar Engineers.

Yang, Y., Chen, J.X., and Huang, S., 2016. Mooring line damping due to low frequency superimposed with wave frequency random line top-end motion. *Ocean Engineering*, 112: 243–252. DOI: 10.1016/j.oceaneng.2015.12.026.

Yuan, Z., Ji, C., and Chen, M., Zhang, Y. 2011. Coupled analysis of floating structures with a new mooring system. In *ASME 2011 30th International Conference on Ocean, Offshore and Arctic Engineering*. pp. 489–496. DOI: 10.1115/OMAE2011-49597.

Yue, Q., Bi, X., Zhang, X., and Karna, T. 2002. Dynamic ice forces caused by crushing failure. In *16th IAHR Symposium on Ice*, Dunedin, New Zealand, December. pp. 2–6.

Yue, Q., Guo, F., and Kärnä, T. 2009. Dynamic ice forces of slender vertical structures due to ice crushing. *Cold Regions Science and Technology*, 56(2–3): 77–83.

Yue, Q., Zhang, X., Bi, X., and Shi, Z. 2001. Measurements and analysis of ice induced steady state vibration. In *Proceedings of the International Conference on Port and Ocean Engineering under Arctic Conditions*.

Zaheer, M.M. and Islam, N. 2008. Fluctuating wind induced response of double hinged articulated loading platform. In *Proceedings of the 27th International Conference on Offshore Mechanics and Arctic Engineering*, pp. 723–731.

Zaheer, M.M. and Islam, N. 2017. Dynamic response of articulated towers under correlated wind and waves. *Ocean Engineering*, 132: 114–125.

Zaheer, M.M. and Islam, N. 2012. Stochastic response of a double hinged articulated leg platform under wind and waves. *Journal of Wind Engineering and Industrial Aerodynamics*, 111: 53–60.

Zhai G.J., Tang, D.Y., and Xiong, H.F. 2011. Numerical simulation of the dynamic behavior of the deep-water semi-submersible platform under wind and waves. *Advanced Materials Research*, 243: 4733–4740.

Zhu, H. and Ou, J. 2011. Dynamic performance of a semi-submersible platform subject to wind and waves. *Journal of Ocean University of China*, 10(2): 127–134.

Index

Note: *Italicized* page numbers refer to figures, **bold** page numbers refer to tables.

225